The Ash Warriors

C. R. Anderegg

Reprinted by
Air Force History and Museums Program
2005

Originally printed by
Office of PACAF History
Hickam Air Force Base, Hawaii
2000

Library of Congress Cataloging-in-Publication Data

Anderegg, C. R.
The ash warriors / C. R. Anderegg.
x, 146 p., [8] p. of plates : ill. (some col.) ; 28 cm. —
Hickam Air Force Base, Hawaii : Office of PACAF History, 2000.
1. United States. Air Force. Air Force, 13th--History. 2. Emergency management.
3. Clark Air Base (Philippines)--History 4. Pinatubo, Mount (Philippines)--Eruption, 1991.
I. Title.

UG634.5.C49 A53 2000
358.4/17/095991 21 2001267523

C. R. Anderegg is a retired USAF fighter pilot.
He lives in Northern Virginia with his wife, Jean.
The Ash Warriors is his first book.

Printed in Hawaii by Valenti Print Group

Foreword

In November 1991 the American flag was lowered for the last time at Clark Air Base in the Philippines. This act brought to an end American military presence in the Philippines that extended back over 90 years. It also represented the final act in a drama that began with the initial rumblings in April of that year of the Mount Pinatubo volcano, located about nine miles to the east of Clark.

The following pages tell the remarkable story of the men and women of the Clark community and their ordeal in planning for and carrying out their evacuation from Clark in face of the impending volcanic activity. It documents the actions of those who remained on the base during the series of Mount Pinatubo's eruptions, and the packing out of the base during the subsequent months. This is the story of the "Ash Warriors," those Air Force men and women who carried out their mission in the face of an incredible series of natural disasters, including volcanic eruption, flood, typhoons, and earthquakes, all of which plagued Clark and the surrounding areas during June and July 1991.

The author of The Ash Warriors knew the situation first hand. Col Dick Anderegg was the vice commander of the 3rd Tactical Fighter Wing when the volcano erupted, and he was at Clark throughout the evacuation and standing down of the base. He brought his own personal experience to bear in writing this story. He also conducted extensive research in the archives of the Pacific Air Forces and Thirteenth Air Force, utilized scores of interviews of those who witnessed and participated in the events, and visited Clark in 1998 to see in person how the installation had changed in the eight years since the Americans left.

This story is one of courage, resourcefulness, and dedication to duty on the part of Air Force men and women called upon to respond to one of the great natural disasters of the Twentieth Century. As the following pages reveal, the Ash Warriors were up to the challenge in every respect.

PATRICK K. GAMBLE, General, USAF
Commander
Pacific Air Forces

For sale by the Superintendent of Documents, U.S. Government Printing Office
Internet: bookstore.gpo.gov Phone: toll free (866) 512-1800; DC area (202) 512-1800
Fax: (202) 512-2250 Mail: Stop SSOP, Washington, DC 20402-0001

ISBN 0-16-050600-X

Preface

After years of encouragement from family and friends to write this book, I discussed it with Richard P. Hallion, the Air Force Historian in March of 1998. He was enthusiastic that the story be told. He was well aware of the accomplishments of the Ash Warriors and knew that their story would be interesting to military and civilian readers alike. During our discussions, I told him what the eruptions looked like from a close-up view. When I finished the story, he said, "Tell it just like that. Don't change a thing."

If the great historian, Stephen Ambrose, is correct, history is the story of those who are leaders and those who are followers. I have tried to make this book just that, the story of the Ash Warriors who faced a killer volcano on the other side of the world. Some were leaders and some were followers, but they all did their duty in the face of terrible conditions and frightening uncertainty.

I served as 3d Tactical Fighter Wing vice commander from July 1, 1990 to August 8, 1991, so many of the observations in this book are from a personal perspective. Where statements of fact or assumption are not attributed to another source, the reader may assume I am speaking from personal experience. Of course, I am solely responsible for any errors of commission or omission. Research for the book was conducted in several areas. I had been collecting surveys from Ash Warriors for several years. Although the surveys were not particularly useful for the book, the network that resulted from the surveys proved invaluable. During April to August 1991, I kept a small notebook from which many events were recalled for the book. After the eruptions, I transcribed the notes onto the only operational computer I could find, and subsequently printed them out as a record of thoughts and events during the crisis.

The PACAF Command Historian's Office at Hickam Air Force Base, Hawaii, provided a wealth of information. They possess about 20 four-inch binders of information about Clark and its closure. The binders contain all situation reports (SITREPS) from both the 3 TFW and JTF-Fiery Vigil as well as the American Embassy in Manila. Without the friendly assistance of Al Miller and Anne Bazzell, I never could have waded through the chaff to find the wheat. After leaving the PACAF office I journeyed to Guam, the present site of Thirteenth Air Force where Deryl Danner helped me immensely by setting up some interviews and leading me through his files. During my visit to Clark, Miss Cefey Yepez set up my whole visit there, and I am deeply indebted to her for showing me the "new" Clark. Thanks to Skip Vanorne and Jim Goodman for providing many personal photographs.

Perhaps the richest vein I mined were personal interviews I conducted, of which there were nearly 100. Everyone I approached wanted to tell me his or her story, and it did not take me long to realize that the interview was a catharsis. People would talk for hours about what they had seen and done. Frequently there were tears and deep emotions bared to the tape recorder. To all those who submitted to my questioning I offer my deepest thanks. You are the Ash Warriors and I did my best to show the world your incredible accomplishment.

Many helped me research and find other Ash Warriors. Col Mike (Air) Jordan was my main source of personnel locators. He and his friends could find anyone, and they frequently spent hours looking for people who had "disappeared." From the US Geological Survey, Chris Newhall and Rick Hoblitt were invaluable. They provided valuable technical advice where I had missed some point in Volcanology 101. No author produces a book without great editors and mine were Pati Wilson and Tim Keck from the PACAF Command Historian's Office. I know there were many moments when Pati repeated the mantra, "patience is a virtue, patience is a virtue" while she tried to repair an old fighter pilot's butchery of English. Additional thanks go to my friends Bob Tone and Helen Puckett who read every word of the manuscript and offered suggestions and corrections.

My last and biggest thanks goes to Jean, my wife, who, while struggling with a serious illness, not only found the time to encourage me, but also to teach all around her the true meanings of courage and love.

Contents

List of Illustrations

Tables

Maps

Photographs

List of Illustrations

Chapter 1 PACIFIC JEWEL

As our world evolved, violent volcanic eruptions formed mountain chains across our planet's face. Throughout pre-historic millennia, as the earth's crust cooled, these enormous mountain ranges settled then erupted again in long cycles of explosion and repose. Many of the world's great mountain ranges were wholly or in part the result of such volcanic activity. During the twentieth century, the most violent of these has been along the infamous "Ring of Fire" that surrounds the Pacific Ocean. This string of mountains runs northward through South America; up through the Cascades; across Alaska; westward into Asia; south through Japan and the Philippines; and ends its journey in the islands of Indonesia. Many mountains in this vast string are active volcanoes that sometimes erupt with spectacular results. Mount Saint Helens, for example, caught the world's eye when it violently exploded in 1980.

In April 1991, Clark Air Base, Republic of the Philippines, was the largest United States Air Force (USAF) base outside the continental United States. The base sprawled over nearly 10,000 acres[1] with its western end nestled in the lush, gently rolling foothills of the Zambales Mountains in western Luzon. Only nine miles west of Clark stood Mount Pinatubo, a small mountain by Ring of Fire standards. Although it stood 5,700 feet above sea level, it barely peeked above neighboring Zambales hills. When viewed from Clark Air Base, Pinatubo seemed insignificant–just another peak in the lush jungle hills. It was a volcano, but it had not erupted for 500 years.[2]

Over 20,000 Americans lived on or near Clark. Most lived on the base, but there were several hundred families living off the base, as well as nearly 4,000 military retirees and their families throughout the surrounding towns. During the Vietnam War, many thousands of USAF members traveled through Clark. Everyone who saw the base, especially those who lived there, marveled at its tropical beauty. There had been many to admire it; the US military had been there for over 90 years.

Base History

During the Spanish-American War in 1898, US cavalry forces operating in central Luzon came across a broad alluvial plain at the foothills of the Zambales Mountains that had good grass for their horses. Throughout their campaign, the Americans were often frustrated at their inability to find suitable fodder for the animals. Thick, strange grass covered most areas of the island. The horses became ill from the unusual grass, and many died. The lush area between the Abacan and Sacobia Rivers, though, had good grass, and the cavalry troops set up a permanent encampment near the barrio Sapang Bato.[3]

During the latter part of 1902, after the encampment had been occupied for nearly a year, the initial boundaries were surveyed and established. On September 1, 1903, an area consisting of 7,669 acres was named Fort Stotsenburg honoring Col John M. Stotsenburg of the 1st Nebraska Volunteers. The colonel had been killed on April 23, 1899, while leading his regiment in action near Quinque, Bulacan, Luzon.

Five years later, President Theodore Roosevelt published an executive order extending the boundaries of the reservation to include a total of 156,204 acres. The first structures were

A DeHavilland D-4 was the first aircraft deployed to Clark Field in 1919.

nipa and bamboo barracks. Because of seasonal heavy rains, wooden frame buildings with corrugated sheet roofs soon replaced these. Officer quarters were built in this style in 1906 and are still in use. By 1913, the first concrete structures had been completed.

Military personnel at Fort Stotsenburg consisted primarily of cavalry regiments until 1919, when an eastern portion of the fort was designated for construction of flight line facilities. The 3rd Aero Squadron of the Army Air Service, including Lt Ira C. Eaker,[*] constructed a half-mile dirt runway, eight hangars, and associated structures. That same year, the facility was named Clark Field after Maj Harold M. Clark, an early aviator killed in a seaplane crash at Miraflores Locks, Panama Canal Zone.

The base's namesake Maj Harold M. Clark, U.S. Army Signal Corp aviator.

On December 8, 1941, Clark Field was devastated by a Japanese surprise air attack. Within 15 days, Fort Stotsenburg and Clark Field were evacuated in the face of overwhelming enemy strength sweeping Luzon from both north and south. The Japanese occupied the installation until 1945, operating from as many as five runways on the base.

American and Philippine forces, commanded by General Douglas MacArthur, reclaimed Clark Field during the battle for Luzon in 1945.[†] Soon after, the base became the home of Thirteenth Air Force, the "Jungle Air Force." The next three years saw an extensive buildup as the bombed-out structures were repaired and new facilities were built. The projects included a runway, hangars, family homes, barracks, power plant, roads, and a perimeter fence.

Following the Second World War, a 1947 military basing agreement granted the United States basing rights rent-free for 99 years. Two years later, Fort Stotsenburg and Clark combined to form Clark Air Base.[4]

The early 1950s saw Clark units move north to assist in the Korean War. A decade later, a new buildup of facilities and personnel at Clark followed the start of combat in Vietnam. Chambers Hall, a bachelor officer housing facility, was built a few years later. Chambers, a high-rise which featured more than 300 rooms, barber shop, restaurant, and other facilities quickly became a prominent landmark. Virtually every new arrival spent time there until housing was available, and thousands of officers on temporary duty (TDY) called it a home-away-from-home.

Clark's more recent history included participation in Project Homecoming (the release of prisoners of war from North Vietnam) and Operation Babylift/New Life, the evacuation of orphans and civilians from Vietnam. Areas around Clark's football field, the Bamboo Bowl, became a tent city to house more than 10,000 refugees.

On January 7, 1979, the Clark reservation became a Philippine base with the signing of amendments to the Philippine Bases Agreement between the United States and the Republic of the Philippines. The base area in which US operations were conducted became a United States facility with a base controlled by the Armed Forces of the Philippines (Clark Air Base Command, or CABCOM) and headed by a Philippine base commander.

Base Layout

Generally, Clark was the shape of a broad triangle laying on its side with the apex pointing west toward Pinatubo. From apex to base, the triangle was over 6 miles wide. Its western end, the apex of the triangle, nestled in the gently rolling foothills of the Zambales Mountains whose lush jungles and rice paddy terraces provided a picturesque backdrop. The Mactan mil-

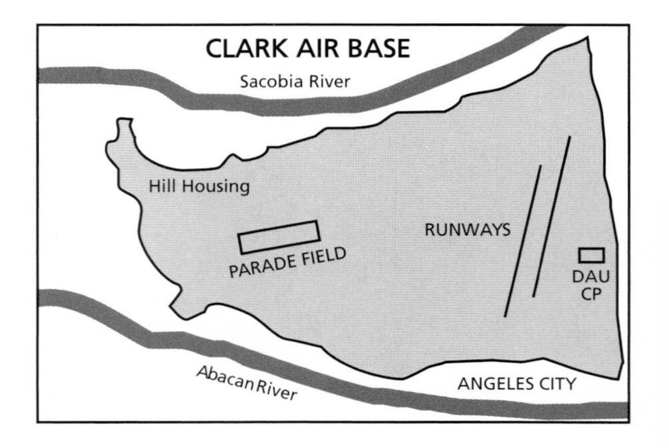

[*] *Eaker became a famous proponent of air power and attained general officer rank.*

[†] *When MacArthur returned and defeated Japanese forces in the Philippines, one of his subordinate commanders was US Army Col Arthur S. Collins, 130th Infantry Regiment commander. On April 20, 1945, Colonel Collins' regiment liberated the city of Baguio high in the mountains north of Clark. A monument on the road to Baguio documents the event. Collins' post there became Camp John Hay. Forty-seven years later, on July 1, 1991, Colonel Collins' son, Lt Col Kevin A. Collins, USAF, commander of Camp John Hay, retired the US flag there and returned the installation to the government of the Philippines. On November 26, 1991, Lieutenant Colonel Collins was promoted to Colonel. For the ceremony, Collins used a set of hand-carved silver eagles that were given to his father in 1944 by the citizens of Baguio as a token of their gratitude.*

itary family housing area comprised most of the west end and was commonly called "The Hill" or "Hill housing." Nearly two thousand homes were on the Hill. Elementary schools, a middle school, a new high school, a shopette (convenience store), and a restaurant complemented Hill housing and gave it a stateside suburbia flavor.

From Hill housing the terrain sloped gently downward to the east and an enormous

Built in the 60's Chambers Hall became a home-away-from-home for the newly arriving bachelor officers.

parade field that was a reminder of Clark's days as a cavalry post early in the century. The parade field, a quarter of a mile long and half as wide, was nearly dead center in the base. The Thirteenth Air Force headquarters building complex stood sentinel at the parade field's west end and was fronted by cannon and flagpoles, which flew American and Philippine flags.

Large homes, called "barns," edged the north side of the parade field. The barns were actually homes built in the early 1900s in a tropical architecture. They were wooden frame homes of one story and stood above the ground on short pilings. Most had large porches, or verandas, on two or three sides. The roofs were steeply pitched and made of corrugated steel sheets. The barns' interiors were equally rustic with *narra,* a local hardwood, floors worn smooth and shining from decades of riding boots, flying boots, bathroom slippers, and gleeful children's sliding stocking feet. These sturdy homes were made for tropical living. The steep roofs protected against torrential monsoon rains while the screened verandas let through breezes during the dry season. The elevation above the ground kept out unwanted jungle critters and provided underneath ventilation to control humidity. The total wooden construction made them flexible in battering typhoon winds that invariably came every summer.[5]

In the early 1900s, the barns were the only kind of housing at Clark, and the US government constructed several hundred over the decades. However, the constant battle with termites, typhoons, and oppressive humidity made maintenance impractical. By the mid-1980s, most of the barns were demolished. In their place, the Air Force constructed nearly 1,000 new military family housing units south and north of the parade field. All told, the base had a little over 3,000 military family housing units. Only a few of the charming old barns remained around the parade field, symbols of Clark's long, proud history that spanned more than 90 years.

The central part of the base also contained the usual assortment of base support facilities: commissary (supermarket), base exchange (department store), automobile service station, movie theater, library, chapels, officers' club, enlisted club, swimming pool, golf course, and a bowling alley. A Baskin-Robbins, within easy walking distance of the parade field and the center of the base, was a touch of America. During warm tropical evenings the ice cream shop did a land-office business as mothers, fathers and children strolled the sidewalk around the parade field, stopping to chat and enjoying the sights of the beautiful base.

Mimosas, bougainvillea, palms of every variety, tropical magnolias, flame trees, countless other flowering plants and shrubs, and sweeping, immaculately manicured lawn areas made the entire base picture-postcard perfect. The most striking features, however, were the enormous acacia trees that lined nearly every street in the housing areas, encircled the parade field, and spread their limbs across enormous spans no mighty oak would ever dare. Many were eighty feet high and twice as wide. Functional as well as beautiful, they provided welcome shade from the dry season's blazing sun.

East of the parade field, the terrain flattened, and the flight line area sprawled across the broad base of the triangle. Two parallel runways ran northeast to southwest. The east runway was new and had opened for operations only a few months earlier. About a mile east of the runway was Clark's east perimeter.

A photo taken in 1998 from in front of the 3rd Support Group headquarters building looking west along the parade fields.

Base operations building overlooking the Clark flight line.

Clark had been enjoying a building boom for several years. In the 1991 fiscal year alone nearly $150 million of construction contracts were awarded.[6] Recent additions to the base included a second runway, airmen's club, 1,000 family houses, three dormitories, non-commissioned officers' club, security police headquarters building, housing management office, flight operations building, officers' club, commissary, and golf course club house. Much of the new construction was aimed at improving the quality of life for Clark's people who were on the other side of the world from "the land of the big BX,*" in an isolated location.

A concrete block wall, eight feet high and twenty-six miles long, enclosed the entire base. The wall was an attempt to keep out economic intruders, a euphemism for thieves, who continually plagued the base.

A complex river system paralleled the flanks of the triangle. Two major rivers, the Sacobia and the Abacan flowed north and south of the base from their headwaters in the Zambales Mountains to the west. Other lesser streams cut through the base. The entire watershed was arid during the dry months, but during the July to November rainy season, the system surged strongly from the Zambales eastward into central Luzon.

* Base exchange.

Jeepneys. The peculiar jeep/car/truck vehicles common in the Philippines that provided public transportation.

Off-Base

Several towns and *barangays* (villages) abutted Clark's imposing wall. The largest one, Angeles City, sprawled south and east. Angeles offered countless bars, nightclubs, and restaurants to serve American servicemen and their families. The area was a beehive of activity, and the streets were clogged with pedestrians, bicycles, and the ubiquitous *jeepney*. The jeepneys were born from post-WW II military surplus Jeeps which locals bought at very low prices and then extensively customized. Jeepneys were ornately decorated and sported gaudy, elaborate paint schemes, plenty of chrome, flashing decorative lights, surrey fringes, and whatever imaginative touches the owners could conjure up. In fact, jeepneys had become an art form; many were decorated with a theme such as "The Terminator" that featured a glowering Arnold Schwarzenegger painted on the side, or "Snow White" that had the seven dwarfs merrily cavorting across the hood.

Most Americans who lived in Angeles lived in guarded compounds; theft was an even bigger problem off the base. Yet many preferred to live off the base because the rental fee for a beautiful home was very low compared to stateside prices. Also, many American servicemen were married to Filipinas, whose families were nearby.

Immediately west of Angeles was the barangay of Sapangbato; to the north of Angeles were Dau and Mabalacat. The local Filipino population numbered about 250,000.

Approximately 10,000 Filipinos worked on the base. About 3,000 were directly employed and paid by the US government. They worked in nearly every office as clerks, secretaries, accountants, and myriad administrative employees. Filipinos were used extensively throughout the base services sector as stable grooms, billeting receptionists, dining room servers, and lifeguards at the three swimming pools.[7]

Many were employed as "local hires" and worked directly for base residents as domestics and gardeners. Since wage scales in the Philippines were low, nearly all the 3,000 families who resided on the base, as well as most of the several thousand families who lived in the surrounding area, employed a "house girl" and "yard boy." Although the gardeners usually came to manicure the landscaping around a home only once or twice a week, it was common for the domestics to live in the home with its American family during the week. Typically, a live-in maid earned $15 a week to do all cooking, baby-sitting, laundry, and light cleaning. Gardeners earned $5-$10 a day to tend the landscaping, wash the cars, and do heavy cleaning in the house. Other local hires rotated through the homes doing manicures, pedicures, and massages. Clever seamstresses earned good livings making tailor-made clothing for American families using pictures from stateside catalogs as patterns. Golf carts were seen rarely on the golf course; nearly everyone used Filipino caddies at $2.50 per round. Most caddies were good players in their own right, so the fee often included free tips and lessons during a round.

Close bonds grew between the American families and Filipino workers, especially between the families and the live-in maids, many of whom had worked on the base for decades. American children grew up on the base with a deep affection for the "extended family member" with whom they lived. Golfers frequently played with the same caddy and established trusting relationships with them.

Many Filipino employees worked their entire adult lives at Clark. One of the best known was Ms Ceferina Yepez. Miss Cefey was the liaison between Clark and the Filipino community. She had worked for the USAF at Clark since the year the Air Force was born–1947. She knew everyone who was anyone in the Filipino community and had her finger on the pulse of American-Filipino relations. An engaging woman with a very quick wit, Miss Cefey arranged social meetings between senior Clark officers and local community leaders with a style and grace admired by all.

Military Operations

In the spring of 1991, Maj Gen William A. Studer was the senior officer at Clark Air Base and commanded Thirteenth Air Force. Thirteenth, with the proud nickname "Jungle Air Force" earned from its exploits in the south Pacific during World War II, was a very small numbered air force. Only the USAF in the Philippines and Guam fell under its command. Although its force strength was small, Thirteenth had a very large area of responsibility that stretched throughout Southeast Asia and included Singapore and Thailand. Clark Air Base and Andersen Air Base on Guam were the only two Air Force bases in the command, although there were several smaller installations such as Camp John Hay and Wallace Air Station in the Philippines which were also in Thirteenth Air Force.

Directly subordinate to Thirteenth Air Force, and the primary unit of Clark Air Base, was the 3d Tactical Fighter Wing (TFW), commanded by Col Jeffrey R. Grime. The 3d Tactical Fighter Wing ran the base and provided all its support functions: housing, logistics, medical and dental care, flight operations, civil engineering, aircraft and vehicle maintenance, supply, recreation, etc. Colonel Grime was also the Clark installation commander and as such was responsible for anything that affected the base or its population. Although General Studer was Grime's boss, Studer was at Clark under the same circumstances as an admiral sailed on a battleship. The battleship (Clark) was the flag ship, but the captain (Grime) commanded the ship.

The primary flying units of the 3 TFW were the 3d Tactical Fighter Squadron, equipped with the F-4E Phantom II multi-role fighter, and the 90th Tactical Fighter Squadron, which was equipped with the F-4G Phantom II Wild Weasel. Three UH-1 Huey helicopters, whose call sign was Cactus, rounded out the wing's flying forces.

In November 1990, the Air Force decided to withdraw all flying units from Clark. By April 1991, only a handful of the 3 TFW's fighter jets remained, and the last of those was flown out June 4.[8]

The 6200th Tactical Fighter Training Group (TFTG) was also under the direct command of Thirteenth Air Force. The 6200 TFTG ran two unique programs: Cope Thunder and Combat Sage. Cope Thunder, which was first held in 1976, was a training exercise that brought units to Clark from throughout the Pacific area to participate in war games. Units from the US Air Force, Navy, and Marine Corps, as well as some foreign air forces, sent fighters, bombers, transports, air-refueling tankers, and command and control aircraft to participate in very intense flying war games.

The Fort Stotsenburg gateposts still stand on the Clark Parade field in 1998. Note the bottom of the post. The date of 1902 is hidden behind the nearly nine inch increase of the ground level as a result of the ash fall.

Cope Thunder was conducted in an area that stretched nearly half the length of Luzon starting at its north end and stretching south nearly 100 miles. Crow Valley, which was only 10 miles north of Clark Air Base, was at the southern end of the exercise airspace and held myriad bombing targets and scoring systems to assess the fighter pilots' bombing skills. In a typical war game, friendly forces, designated "Blue," would take off from Clark, fly up central Luzon, then fly south through the exercise airspace to attack the targets in Crow Valley. Enemy forces, designated "Red," defended Crow Valley. The Red forces consisted of "enemy" fighters, frequently US Navy jets, and an impressive array of radar transmitters that replicated Soviet surface-to-air (SAM) missiles and radar-controlled anti-aircraft artillery (AAA). All Red defenses had the capability to score their "shots" against Blue attackers. Every morning and every afternoon, throughout the year, dozens of Blue fighters and support forces hurtled south into Crow Valley, honing their wartime skills in thunderous, swirling battles. To understand the scope of these exercises one need only imagine what the movie "Top Gun" would be like had Cecil B. Demille produced it. Cope Thunder, by most accounts, provided aircrews with the best training available in the Pacific.

The 6200 TFTG's other mission, Combat Sage, was every bit as valuable and challenging as Cope Thunder. Combat Sage's mission was to evaluate the Pacific Air Forces' (PACAF) fighter units' capabilities to fire air-to-air missiles. Fighter units from PACAF came to Clark regularly to fire missiles at target drones launched from Wallace Air Station, a small installation north of Clark. The Combat Sage troops provided missiles that were equipped with telemetry packages that transmitted such data as missile performance, pilot performance and miss distance to Combat Sage telemetry gathering stations. The data was then used to analyze missile, aircraft, and pilot performance. Commanders, logisticians, and weapons developers throughout the USAF used the data to improve aircraft and missile systems as well as pilot training programs.

3d Wing F-4 parked in front of the Clark Control Tower.

Several associate units (units not under direct operational command of 3 TFW) resided on the base as well. The largest of these was the 353d Special Operations Wing (SOW), commanded by Col Leon E. Hess. The 353 SOW was equipped with MC-130E Combat Talon aircraft as well as MH-53, Pave Low helicopters.

Clark was an ideal location for the SOW; the terrain of Luzon offered a variety of features that ranged from broad expanses of rice paddies to high mountains. For pilots who might be tasked to fly special operations missions anywhere in Asia, Luzon was a superb training ground. Since much of the SOW's training was done at night, Luzon's thinly populated expanse permitted low level flight with very little disturbance.

Over the years special operations forces had earned the moniker "snake eaters," because special operations forces often worked behind enemy lines and had to live off the land. The SOW at Clark was a highly trained and professional unit that carried the "snake eater" image proudly.

Over the years, the infrastructure of Clark evolved toward self-sufficiency. Electrical power was generated by two large electrical production facilities. All water came from deep wells within the base boundary. Water from these wells was pumped to several storage tanks strategically placed across the base. There was even a dairy which produced milk and milk products so the supplies were fresh for consumers. Television and radio stations broadcast American TV shows, including sports and news, to work places and homes both on and off the base. A weekly base newspaper, the *Philippine Flyer*, covered local events, sports, and public service announcements.

I Survived . . . .

By the spring of 1991, however, Clark's people were tired and stressed. Over the previous year, a nearly unimaginable series of events seemed to pile one on another. In May 1990, two airmen were assassinated near the base. The killers were members of the New People's Army (NPA), a communist-based guerrilla group which strongly opposed a US presence in the Philippines. Other NPA assassinations had happened over the years, and the loss of these two airmen only added to the uneasiness all Americans felt while travelling in the area. In response to these most recent killings, General Studer ordered Americans confined to the base. Those who lived off base had to stay in quarters and could only leave to drive to and from work. The restrictions only lasted a short time, but thereafter, whenever intelligence indicated another possible threat to American lives, the "lock down" would be reinstated. In July 1990, a tremendous earthquake, measuring 7.8 on the Richter Scale, shook central Luzon.[10] Although little damage occurred on Clark, it was a frightening experience for all. Many airmen were involved in relief operations in the hardest hit provinces north and east of Clark. It was not a pretty sight, especially in Cabantuan, where a poorly constructed school collapsed on 250 children. The installation at Baguio, Camp John Hay, was badly damaged, and the town of Baguio had hundreds dead. Clark airmen were heroic in their relief efforts, although the horror of dragging mangled bodies from crumpled buildings took its toll.

One beautiful morning in early April, base residents could see a tall, puffy, white plume rising a thousand feet into the bright, blue sky

In the fall, not long after the earthquake, the local Filipino workers' union voted to strike, and nearly 3,000 workers on the US government payroll walked out. It was a bitter strike, and again Americans were locked up because of the threats of violence to them by the strikers. The strike lasted only ten days, though, because downtown businessmen put pressure on the strikers. The lock down of Americans had caused a drastic reduction in business.[11]

If these pressures were not enough, several typhoons swept through the area, a typical occurrence during the late fall and early winter. Many remembered a super typhoon that smashed through the base in 1989. These huge storms caused massive flooding throughout Luzon. All in all, to maintain that the mood and morale at Clark were not high was a classic understatement.[12] In a display of unmistakable gallows humor, many wore T-shirts which proclaimed in bold letters, "I survived . . . ." These words were followed by a long checklist of frightening and dangerous events: typhoons, assassinations, floods, strikes and coup attempts against the Philippine government of President Corazon C. Aquino.

Base Negotiations

Despite the continuing string of man-made and natural disasters, the one series of events which most disturbed the Americans at Clark were the base agreements negotiations. The long series of meetings, which lasted from the fall of 1990 into the summer of 1991, were bitter and contentious.[13] The US team, headed by Ambassador Richard Armitage, had as its purpose of the negotiations to establish a new agreement to replace the one that would expire in 1991.

Although the Americans at Clark thought their beautiful base was essential to continued US presence in the region, some national leaders in Washington, DC expressed concern with a perceived "lack of leadership" in the Aquino government and the "corruption that was endemic in her family."[14] Further, Aquino and her government had failed to spend millions in US aid. Then, adding insult to injury, she had refused a meeting with Secretary of Defense Dick Cheney despite the fact that US jets from Clark had supported her during a recent coup attempt.[15] When Armitage met with Secretary of State Baker prior to the first round of negotiations, Baker told Armitage he did not care if an agreement was reached, nor did President Bush. However, Bush wanted other Asian allies to see that the United States was interested in the Pacific rim. Armitage was given instructions to negotiate

A security officer looks on as strikers picket the main gate.

honorably and fully, but if no agreement resulted, so be it. He was told, "do your best, but we don't care."[16]

The first six months of the 1991 negotiations were trying. At the beginning of the year, the Philippines wanted guaranteed fixed "rental" for the term of any agreement reached, whereas the United States only wanted to pledge a "best effort" to adjust the annual fee in accordance with other factors, primarily the utility of the base. In mid January, US negotiators agreed to return Wallace Air Station to the Philippine government, an act which the anti-American Philippine press saw as "conciliatory." Much attention, both in the media and among supporters, turned to a limited access to Clark, whereby only the Special Operations Wing, Cope Thunder and Combat Sage would use Clark. Many Philippine voices, who claimed to be pro-bases supported only a five-year extension for access at a rental fee of $500 million annually.

It was difficult if not impossible for Armitage and the US team to pin down exactly what the Philippines and its chief negotiator, Raul Manglapus, wanted. Aquino stayed clear of the negotiations, and the resulting lack of leadership permitted a cacophony of factional clamoring for a variety of schemes. By the end of January 1991, these posturings seemed nothing more than a disorganized effort to jack up the price of an agreement. The major sticking point was compensation. The US Joint Chiefs of Staff considered anything less than a ten-year agreement to be temporary; therefore, compensation should be less. Armitage and his team, however, seemed unable to move Manglapus off his position of a seven-year agreement at a fixed price.

Negotiations continued through the spring of 1991 by fits and starts, with issues rising and falling in concert with Philippine demands affecting national sovereignty, any of which could be resolved were the United States to offer "enough" money. The sixth round ended with no progress, as Manglapus insisted on $825 million a year for a maximum of seven years. Armitage held firm that the maximum the United States would pay was $360 million a year for ten years; any shorter term would mean a significant reduction in compensation. The collapse of the talks seemed imminent.[17]

All of this was even more confusing to Clark's residents and the surrounding American and Philippine community. The constant yo-yoing of the talks created an atmosphere of uncertainty about their futures. The local Filipinos and the American servicemen and families were a tightly knit community with many close ties. To most, the idea that the base would close was nearly unthinkable. Further, each time Armitage and Manglapus entered into a new round of discussions, the NPA threatened to assassinate more service members as a signal of their desire to see the Americans out of the Philippines. In response, General Studer directed a "lock down" of the base prior to each round. Those who lived on base could not leave the base, and those who lived off-base could only travel direct, protected routes to and from work. General Studer full well knew the impact of the lock downs, but he was determined to do everything in his power to protect his people.[18] Ultimately, a siege mentality with its attendant cabin fever set in throughout the base.

The Last Straw

One beautiful morning in early April, base residents could see a tall, puffy, white plume rising a thousand feet into the bright, blue sky over the Zambales Mountains. For those few who know the area well, it was clear that the plume was emanating from Mount Pinatubo. It was April 2. The next ten weeks would prove it to be no belated April Fools' joke.

NOTES TO CHAPTER 1

1 Briefing, 3 TFW/CC to PACAF/CC November 14, 1991, PACAF/HO, Hickam AFB, HI.

2 Christopher G. Newhall, ed and Raymundo S. Punongbayan, Fire and Mud (Seattle: University of Washington Press), 1996, p 3 [hereafter cited as Fire and Mud].

3 United States Air Force Fact Sheet, "Clark Air Base, Republic of the Philippines," 3 TFW/PAO, March 1984.

4 Briefing, "Disaster in the Philippines," n.d., on file PACAF/HO, Hickam AFB, HI.

5 C. R. Anderegg, personal journal, 1991 [hereafter cited as author's journal].

6 Interview, Col James E. Goodman with the author, November 11, 1998 [hereafter cited as Goodman interview].

7 Interview, Ceferina Yepez with author at Clark AB, Philippines, October 30, 1998.

8 Pilot's log book, C. R. Anderegg [hereafter cited as Anderegg, pilot log book], entry from June 4, 1991.

9 History, Ash Warriors: The relocation of the 353d Special Operations Wing, June-December 1991, Maj Forrest L. Marion, HQ Air Force Special Operations Command/History Office, p.3.

10 Fire and Mud, p 5.

11 Interview, Lt Col Ronald Rand, with PACAF/HO, September 14, 1992 [hereafter cited as Rand interview].

12 Ibid.

13 Pacific Air Forces Annual History, 1991, PACAF/HO, Hickam AFB, HI, p 68 [hereafter cited as PACAF 1991 History].

14 Interview, Ambassador Richard Armitage with the author, January 27, 1999 [hereafter cited as Armitage interview].

15 Ibid.

16 Ibid.

17 PACAF 1991 History, p 69-70.

18 Interview, Maj General (Ret) William A. Studer with the author, January 5, 1999, tape on file PACAF/HO, Hickam AFB, HI [hereafter cited as Studer interview].

Chapter 2 VOLCANO? WHAT VOLCANO?

A few days after the steam plume appeared, the phenomenon was discussed at the daily 3 TFW "stand-up," or senior staff meeting. This meeting, which occurred in similar fashion at every Air Force base around the world on a daily basis, was chaired by the installation commander and included his staff and representatives of every organization on the base. The wing commander's principal deputy, the vice wing commander, chaired the meeting on occasions when the commander was absent. Such was the case at this meeting because the wing commander, Col Jeff Grime, was flying an F-4 mission over the Crow Valley range.

The first order of business was to hear a briefing that reviewed all base activities from the previous day such as flying results, maintenance status, supply status, and hospital activities. A briefing that detailed upcoming events followed. During the last ten minutes of stand-up, each staff member had the opportunity to speak individually. When it came to Col John Murphy, the 3d Support Group Commander, he surprised everyone by saying that the steam rising out of the Zambales Mountains was from a volcano. In fact, the volcano was Mount Pinatubo.

The vice wing commander, Col Dick Anderegg, responded, "Volcano? What volcano?" This remark expressed the immediate reaction of nearly everyone in the room and foreshadowed the reaction nearly everyone on the base would have as word spread about the volcano.

Colonel Murphy, as support group commander, oversaw all the base's support functions. An eager, very intelligent officer, Colonel Murphy was highly regarded throughout the base as an officer who was a "people person," one who took great pains to look out for his troops. He had good ideas and was a ball of fire putting them into action. On one occasion, after he learned that retirees, who numbered some 4,000 living near the base, were unable to purchase adequate provisions at the base commissary and base exchange, he quickly resolved their complaint.[1]

During the previous few years, a Filipino company, trying to capitalize on speculation that the area around Pinatubo was rich with geo-thermal energy, drilled some deep wells. The purpose of the wells was to tap into what the speculators believed to be large stores of steam trapped underground. The wells might produce steam power that could drive electrical generating stations. The gamble did not pay off, however, and the company abandoned the effort.[2]

Colonel Murphy heard that the steam plume was coming from an abandoned well that had broken open. He investigated further and learned that the wells had nothing to do with the steam. In fact, a small explosion had occurred on the northwest flank of Pinatubo. The source of his information was the Philippine Institute of Volcanology and Seismology, or PHIVOLCS (pronounced Fee-volks).

Questions flew like arrows at a circled wagon train. Where? How? When? Big? Small? Colonel Murphy had no answers to these questions. As the officers left the large briefing room every conversation was similar. A volcano? How could there be a volcano there? It was just a small mountain range with a few large hills. It sure didn't look like a place for a volcano!

Steam rises from vents along the northwest flank of Mount Pinatubo in June, 1991.

The Scientific Wheels Start Turning

In Manila, less than sixty miles southwest of Clark, Dr. Raymundo Punongbayan, Director of PHIVOLCS, was well aware that Mount Pinatubo was a volcano. Only days before, Sister Emma, a nun who worked with Aeta tribesmen living on the flanks of Pinatubo, came unannounced into Dr. Punongbayan's office. She told him there had been a large explosion near the mountain's summit and a lot of ash was on the ground. The air was laced with a strong sulfur smell. Dr. Punongbayan immediately dispatched a PHIVOLCS team to investigate. The team returned the next day to report that a fissure, approximately half a mile long, had opened high on the northwest flank of the mountain. The blast killed a large area of vegetation. Further, they reported steam and sulfur fumes were emitting continuously from the fault.[3]

Punongbayan sent his team back to the mountain, this time with a seismograph to determine if any subterranean activity was occurring, and he did one other thing. It turned out to be an act that ultimately saved thousands of lives. He called his old friend, Dr. Chris Newhall, at the US Geological Survey (USGS) in the United States.[4]

During the middle weeks of April, most base residents took some time to peer curiously at the column of steam rising out of the Zambales west of the base. The best view was from the parade field where the columns seemed to rise out of Thirteenth Air Force headquarters flanked by the US and Philippine flags. It was the dry season, during the northeast monsoon, so the skies were perfect blue, and the columns of steam that rose against them were startlingly white and quite majestic. Even though it was beautiful to see, those living on the hill, near the west wall, could smell clearly the sulfur emissions from the vents, and Colonel Grime was concerned the residents' health could be affected. The hospital commander, Col Brian Duffy, suggested one of his teams investigate. The team, which had been trained to test for a variety of poison gases as part of their military training, took air samples for several days on the hill. They determined that, despite the occasional repugnant odor, there was no cause for concern. Colonel Grime accepted their findings but directed continuing checks. Military training was starting to kick in: plan for the best; expect the worst; accept anything in between.

Meanwhile, Dr. Punongbayan at PHIVOLCS and Dr. Newhall at USGS were trying to get a USGS team to the Philippines. PHIVOLCS was undermanned because several scientists were abroad working on advanced degrees, and the Taal volcano not far from Pinatubo was acting up, too. Dr. Punongbayan requested assistance through the US Agency for International Development (USAID), but they refused.

Dr. Newhall pressed the issue by calling directly to Clark and speaking with Col Bruce Freeman, vice commander of Thirteenth Air Force. Dr. Newhall explained to Colonel Freeman that the Air Force needed USGS at Clark, but USAID did not see it that way. Newhall recalls Colonel Freeman's reaction as "expletives deleted."[5] The colonel went directly to the American Embassy in Manila to argue the case, thus completing an "end run" around USAID.

The next day, a surprised Dr. Newhall received a terse fax from USAID saying that they expected his team to be on the next plane out. Although they were allocated a mere $20,000, the fax, with pointed sarcasm, suggested they "should not feel obligated to spend all of it."

Dr. Newhall, along with volcanologists Andy Lockhart and John Power, got on the first available airplane. They took 35 trunks of gear on a flight halfway around the world. The $20,000 was gone before they got to Clark.[6]

Dr. Andy Lockhart, geologist from the US Geological Survey Center, West Virginia, monitored recording instruments during Mount Pinatubo's upheavals.

The small USGS team arrived at Clark April 24. Volcanologists John Ewert, Rick Hoblitt, and Dave Harlow were not far behind. Murphy set them up in a two-story, four-unit, base house on Maryland Avenue where they could monitor their equipment and could see the volcano from the second floor windows. The living room became a makeshift conference room. Bedrooms on either side of the main unit provided sleeping quarters for the small crowd of the USGS-PHIVOLCS team. They named the house PVO (Pinatubo Volcano Observatory). PHIVOLCS added people to the team, and soon all were busy working to find out what Pinatubo was up to. PVO, with its air-conditioning, kitchen, bathrooms, and showers was a good working environment. One volcanologist joked that it was better than the South American chicken coop in which he had last worked.[7]

Seismometers emerged from the 35 trunks of equipment. These sensors are about the size of a very large thermos bottle and powered by a car battery. Their purpose is to feel the vibrations of earthquakes. The volcanologists selected seven sites around Pinatubo ranging from one to 17 kilometers from its summit. At each of these locations, a seismometer was installed. The process went very quickly because base helicopters, call sign Cactus, airlifted the scientists and their equipment to the installation sites. To install the sensors, workers dug a hole in firm ground, bedrock if they could find it, then partially filled the hole with quick-setting cement. The sensor was then securely fastened to the hardened concrete and attached to a common car battery for power. A transmitting antenna stuck out the top. One of the sensors was installed on the base as well. Three radio relay sites were installed to insure all data could be transmitted to PVO.[8]

As soon as the Clark sensor was operational, it confirmed what Dr. Punongbayan's initial teams had found. Pinatubo was very restless. Deep below the surface, magma was expanding. The expanding magma cracked its overarching layers of rock as it pushed upward. Small earthquakes, which the seismometers could feel, signaled the magma's expansion. Although PHIVOLCS' sensors were good, those that USGS brought in the trunks were state-of-the-art equipment. The signals they sent back were displayed on large rotating drums covered with graph paper. When the sensor felt an earthquake, the ink needle on the drum squiggled like a lie detector machine. The room in PVO where these drums were located quickly became known as the drum room. There was always a volcanologist on duty watching the needles squiggle back and forth on the slowly rotating drums. Visitors to the drum room quickly learned the protocol. Interested drum observers stood a foot back from the drums, hands in pockets, bent forward at the waist and peered down at the squiggles.

When some people think of earthquakes, they envision San Francisco falling down. Such was not the case in the early days of Pinatubo's activity. The earthquakes, which the drums displayed, could not be felt at the American base. They were very small and very deep below the surface.

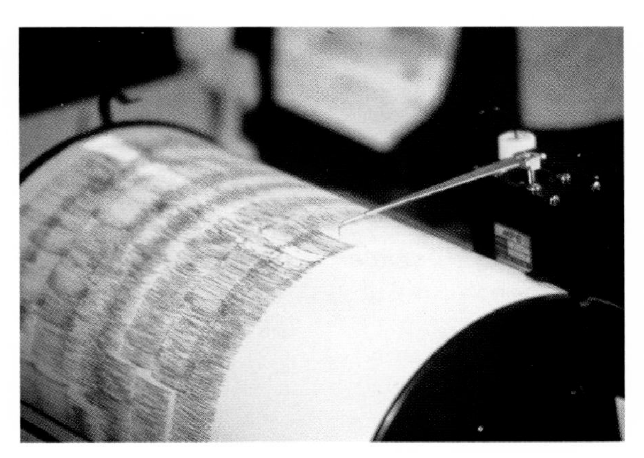

A seismograph records tremors from Mount Pinatubo.

Lockhart and Power seemed to be scientists who could wire anything together and make it sing Dixie. The drum room was a maze of wires and cables. At the confluence of two particularly jumbled messes were linked laptop computers. A software program, called RSAM (relative seismic amplitude measurement), was able to measure a predicted energy level within Pinatubo. Even those with little scientific background could understand the RSAM's story; as the squiggly line on the computer screen rose higher, so rose higher the energy building in the mountain. As each day passed, the squiggly line climbed higher.

Another part of the software could display a graphic image that showed not only the location of each small earthquake, but its depth as well. Every time one of the seismometers registered a quake, a small, green + would show on the monitor. As the earthquakes accumulated, it became easy to see the path of building pressure under the mountain.

Not everything went smoothly as Col John Murphy and Lt Col Al Shirley settled the volcano team and provided for their needs. When Dr. Newhall tried to buy 20 car batteries at the base service station, the attendant told him, "Sorry, one per customer per car per year." Batteries were a valuable commodity on the black market that flourished outside Clark's wall, so a limit had been placed on their purchase. Colonel Murphy intervened with his counterpart who ran the service station concession. Newhall got his batteries, and the sensors started transmitting. Dr. Newhall, having long since spent the pittance provided by USAID, paid for the batteries with his personal credit card.[9]

Several USGS team members sported beards. Of course, this team of scientists looked very much out of place with their beards and very casual clothing sprinkled into Clark's military environment of squeaky-clean uniforms and short haircuts. It did not take long for someone to start calling the USGS team "the beards." Nicknames are common in fighter units, and Clark was, first and foremost, a fighter base. Nicknames are not given to fighter pilots until they become fully mission-ready and part of the fighting team. The USGS scientists became part of the team quickly, and just as quickly, their nickname became used widely.

Gradually, the team that monitored the mountain emerged. Colonel Murphy (nicknamed Whopper because of his support responsibilities, "Have it your way") provided housing, transportation, and other needs to the USGS. Dr. Newhall headed the USGS team, served as liaison with PHIVOLCS, and educated Clark's senior leadership. His lectures, or Geology 101 as some called them, were given many times to many groups. Maj Gen William A. Studer, the Thirteenth Air Force Commander, kept the embassy in Manila and the Pacific Air Forces headquarters informed about what was happening at Clark. Colonel Grime made final decisions about any actions to be taken in response to the volcano's threat. Colonel Grime's vice wing commander, Colonel Anderegg, would be the director of the crisis action team (CAT) which would execute Grime's orders.

On the slopes of Mount Pinatubo, Dr. Chris Newhall, one of the USGS volcanologists studies the mountain.

Right at the start, a philosophical difference emerged between Colonels Grime and Murphy. Colonel Murphy wanted to flood the base with information about the volcano with an appearance by USGS team members on television and a series of *Philippine Flyer* newspaper articles. His logic was that "well informed was well armed." Colonel Grime took the opposite view, though. He did not want any information put out that was not absolutely and certainly accurate. His logic was that incomplete information, or, worse yet, wrong information would only serve to frighten people needlessly. As events progressed, the wing commander and support group commander discussed and argued their views on this important issue more than once. Colonel Grime's point of view, as the officer ultimately responsible for the safety of his people, prevailed. The policy was for no information to be disseminated unless it became necessary, and there was no chance of the information being reversed in the future. Grime knew that if the volcano progressed to an eruption he could only cry wolf once.[10]

Lt Col Ron Rand (call sign "Mercury," the winged messenger) was the base public affairs officer. He worked for General Studer at Thirteenth Air Force but had responsibility for information dissemination across the base. He became the single point of contact for information about the volcano. Americans on and near Clark were very familiar with Rand's face and voice. During the incredible string of hardships the base experienced, his was the face everyone saw on Far East Network (FEN) television, and his was the voice on the radio. So it would continue during the weeks as Mount Pinatubo cooked. When Colonel Rand spoke on television, he delivered his message with a total deadpan face and spoke with what could only be described as a flat monotone. Clark residents and workers dubbed Rand "Colonel Bland."

By mid-May all the scientific gear at PVO was operational, including two tilt meters that had been installed high on Pinatubo's flanks. The job of the tilt meters was to determine if the ground started to swell or shift, an event that might presage an eruption. USGS team member John Ewert was on-board by then, and the tilt meters were his babies. Although Ewert was beardless and did not wear flip-flop sandals, his prognostication skills proved to be nearly visionary. He was right so often about what the mountain would do, some suggested he should be sent to Las Vegas with a month of everyone's pay. It would have been a wise investment.

Also on-board was the team's primary rock hound, Rick Hoblitt. A slight, wiry man with a full beard and a quick smile, Hoblitt arrived from the United States with a severe toothache. Murphy immediately saw him off to the base dental clinic where the dental clinic commander Col Charley Dufort treated the ailing volcanologist.[11] Despite the pain and jet lag, Hoblitt went to work immediately surveying the area for signs of Pinatubo's past. Base helicopters frequently took Hoblitt and Newhall, who was also a geologist, out to the riverbeds and other points of geological interest to gather rocks. Hoblitt did some on-the-spot analysis for a "quick-look," then sent the rocks on to USGS laboratories in the States.

As the scientific evidence accumulated, the base's senior officers continued to get the Geology 101 course from Newhall and the others. The central computers at PVO were programmed to locate the earthquakes from the sensors' data transmissions by triangulating each earthquake's position and depth. They also measured the amplitude (relative strength) of each earthquake and its frequency (pitch).

Pinatubo was turning out to be a Plinian, or explosive volcano

This information was critical. Position and depth were important to determine if and where magma was moving upward, toward the surface. The significance of the amplitude or strength of the earthquakes was a measure of the force of the magma as it fractured rock in the earth's crust. The frequency or pitch thus would provide an indicator of what was causing the earthquake. The sharp cracking of rock at a high frequency would indicate that magma was forcing its way through solid rock. Lower frequency earthquakes would often occur as the magma got closer to the surface. As volcanoes progressed toward an eruption over a period of weeks, months, or even years, the earthquakes came in swarms, ever increasing in strength and moving toward the surface. As the magma expanded and moved, it caused earthquakes of notably higher strength and lower frequency.

As the data grew, Dr. Newhall was able to express his evaluation of Pinatubo. The news was not good. It appeared as though it was not a hydrothermal event – one that merely boils subsurface water and throws it into the air as harmless steam like Old Faithful. Nor was it a lava-producing volcano similar to those people ooh and aah over in Hawaii where the lava flows in majestic ribbons. As Rick Hoblitt said, Pinatubo was not "the kind where you sit on your front porch, sip a cocktail, and wonder if it will get to your property within the next four generations." No. Pinatubo was turning out to be a Plinian,[*] or explosive volcano. Just like Mount Saint Helens.

The geological survey in progress by Mr. Hoblitt gave a glimpse of the volcano's history. The three most recent major eruptions occurred approximately 500, 3,000, and 5,500 years ago. "The general pattern of eruptions was the same: episodes of voluminous explosive eruptions separated by centuries to millennia of repose."[12] Many good examples of pyroclastic flows showed deposits in the riverbeds both north and south of the base.

At this point in the training session, the vice wing commander interrupted. "What is pyroplastic?"[†] Dr. Newhall, ever the educator, responded, "Not pyroplastic. The word is pyroclastic. Hot fragments. Pyroclastic flow is the main danger of a volcano such as Pinatubo. If the eruptions are big enough, the huge amount of ash ejected rolls down the volcano's flanks in enormous clouds. These flows typically are 900 to 1,000 degrees centigrade, move at between 50 to 100 mph, and because they are so dense, quite simply wipe out everything in their path." This last comment elicited only silence.

Newhall added, "We have found evidence that the last eruptions deposited pyroclastic material in the Sacobia River on the north side of the base and the Abacan River on the south side. Of course, those rivers come very close to base housing. Perhaps a few hundred yards. The real problem is that they can jump the banks if they are big enough."

The vice wing commander responded, "So what you're telling me is that *if* the mountain erupts, and *if* the eruptions are big enough, and *if* pyroclastic flows come down the watershed, and *if* they jump the banks, the people who live in those homes are toast. Right?"

He answered, "Yes, that could happen, based on the type of volcano we know Pinatubo to be, and what has happened in the past which is frequently a good indicator of the future."

In a journal kept on his personal computer, Anderegg wrote about his thoughts at that moment. "I had no idea how to deal with this information on either a personal or professional level. What they had just said scared me, really scared me. Not so much for myself or any immediate danger. But how would we manage this thing? How would we make decisions? Thousands depended on us to make the right call."

** Pliny the Elder died at the explosive eruption of Mount Vesuvius.*

† Hot/fiery fragments is the meaning of pyroclastic, from the Greek.

"I had no doubt that if the volcano continued to develop, Grime was going to be in a very tight situation, and we didn't know anything about volcanoes, especially this one. We were military men. Ask us about Soviet tactics, logistics, morale, equipment. Ask us what it's like to pull six times your own gravity in a hard turn while you're trying to maneuver for a kill in a dogfight. Ask us how to motivate subordinates to achieve their best. But volcanoes?"[13]

Colonel Grime and Colonel Murphy were still of a different mind about planning for a possible evacuation as well as what information to pass to the troops. Colonel Grime was very concerned that passing information to people piecemeal was a recipe for disaster. He did not want panic to spread throughout the base. Colonel Murphy's desire was to get everything that was happening out in the open and let people make plans.

By the middle of May, even the volcanologists were unwilling to determine which way the volcano might go. They were sticking to "a 50 percent possibility of an eruption within a year" status.[14] Colonel Grime saw no reason to react based on such flimsy evidence. His efforts to force USGS into being specific in their prognostications were unsuccessful. He wanted them to give him a series of decision points from which he could react, but they would not do so.[15] The USGS business is to give people at home and abroad a heads-up about volcanic business, but the volcano had not started down the road to an inevitable eruption, yet.[16]

Colonel Grime's attempts to keep a lid on rumors and fear were not working. When the USGS sent and received a fax, it went through the base weather office. This was akin to posting the faxes on the base bulletin board since the messages went through so many hands long before they reached their intended destinations. In order to fill this information leak, which was only one of many, the USGS got their own fax machine at PVO.[17] Freeman, who had gone out on a limb to get the USGS team to Clark, was also concerned about rumors and asked scientists not to speculate about the volcano with office workers when they called.[18]

Steam rose from the vents along the northwest flank of Mount Pinatubo in June, 1991.

Colonel Murphy, in an attempt to follow the wing commander's guidelines while preparing for the worst, began clandestine preparations for an evacuation.[19] Lt Col Jim White was the commander of the central base personnel office, the chief human resources officer on the base. Murphy trusted him and late one evening in early May called White to a meeting at Murphy's house. At this meeting, Murphy tasked White to begin planning for an evacuation. One of the first tasks, according to Colonel Murphy, was to find a place from which a security force could operate which was well away from the base. Murphy was not convinced that evacuating to the far side of the runway was far enough away for safety.

The two officers started driving around the local area east of the base looking for a suitable location. They found the Pampanga Agricultural College ten miles further east of the base, and discussed with the president of the college their desire to store food and water at the college. The president did not seem concerned about why they wanted to do so, but he agreed after learning that Murphy was willing to have the base civil engineers make some minor improvements to the college. Such projects were not unusual. The base had a very active civic actions program that frequently remodeled schools and sent medical/dental teams into primitive areas to administer to the poor. Shortly thereafter, Col Jim Goodman, the base civil engineer, sent workers to make a few thousand dollars worth of repairs, and the supplies were squirreled away in two storage rooms.

Colonel Murphy's intuition was that an escape route was needed to the east for security forces that would be left behind to guard the base in the event of a large-scale evacuation. No such plan had been discussed officially, but Murphy acted on his intuition and put his deputy, Jim White, to work.

The only disconcerting aspect of the agricultural college was that it was directly on the flank of a dormant volcano, Mount Arayat, which rose ominously from the rice paddies east of the base. Arayat, which was some 2,700 feet higher than the surrounding elevation, was familiar to anyone who visited or lived at Clark. Many who heard about the volcano near Clark immediately assumed it was Arayat. It was not.

Arayat remained placid. Pinatubo, a small, smoking hill, its top barely visible above the surrounding jungle eight miles west of Clark, was the bad boy.

In PVO, the drums kept turning. Earthquakes under the volcano started to come in swarms, although they still could not be felt on the base. The gathering of green + marks on the graphic depiction screen was getting thicker and thicker. A clear pattern was starting to develop. A "chimney" was forming vertically from several miles below the surface. The chimney pointed directly upwards toward a small fault line that was the source of the open steam vents on Pinatubo's northwest flank.

By May 20, Murphy and White had given considerable thought to how an evacuation might be conducted, and Colonel Grime was ready to start serious planning for an evacuation.[20]

The real question was how far would people have to go to be safe

Colonel Grime met with his vice wing commander, Colonel Anderegg, and gave his instructions for initial contingency planning to start. He wanted to structure emergency responses to the volcano similar to emergency actions with which the base populace was already familiar. He instructed the vice commander to form a crisis action team from the principal commanders across the base. The goals were simple and straightforward. First, and foremost, there would be no loss of life. Military members, their families, and other base employees must be kept out of harm's way at any cost. Second, every plan must assume that the worst could happen. Third, all military and personal property must be protected, but protection of property was to take a complete back seat to keeping people safe.[21] Further, the vice wing commander would run emergency actions from the command post, and the wing commander would stay mobile around the base so he would know what people were thinking and doing. He emphasized that extreme care must be exercised when passing information to the general public. Nothing was to be said of an official nature unless Colonel Rand was saying it. Colonel Anderegg asked when Colonel Murphy could put the USGS team on television to educate people directly. Colonel Grime replied, "Not yet." He was still not convinced such a display would not cause more problems than it would solve.[22]

At this point, everyone, including the USGS-PHIVOLCS team, was unsure what course the volcano might take. At any moment, the pressure building inside the mountain could reach an area of over-arching rock that was stronger than the pressure. If so, the activity would cease. Simply stated, the volcano might go back to sleep.

Colonel Grime was especially skeptical that the volcano would continue towards an explosion.[23] The volcanologists freely admitted that Pinatubo, like all volcanoes, was difficult to predict. They stressed in their educational discussions that everyone had to view the volcanic activity in geological rather than calendar time. The few weeks of Pinatubo's activity were a nano-second in geological terms. The activity could go on for years before resolution.

Colonel Grime insisted that information about any plans be kept low-key and there be no speculation from anyone credible about what might happen with the volcano. Information about the volcano would be passed through the chain of command so that it came from him to the group commanders, to the squadron commanders, to the first sergeants, and then to the troops. The wing commander could make, and would make, hard decisions when needed. However, he insisted on having complete information before deciding.

Typically, the duty of crisis action team commander falls to the installation's host vice wing commander, in this case Colonel Anderegg. Usually, the CAT's membership is composed of essential leaders from all organizations on the base. Following his meeting with Colonel Grime, the 3d TFW vice commander called a meeting with those who would comprise the CAT should the volcano's activity reach a level that threatened the base.

The author, Dick Anderegg, gives instructions to members of the team.

This first planning meeting assembled in the 3 TFW conference room and, in a meeting that lasted nearly two hours, laid the groundwork for everything to follow. The officers gathered around the table represented every major area of the base. Col John Murphy--support group; Col Dave Ray--flight operations; Lt Col Hank Camacho--aircraft maintenance; Col Hal Garland--logistics; Col Jim Goodman--civil engineering; Col Bill Dassler--security police; Col Al Garcia--communications; Col Brian Duffy--hospital; Col Bill Meeboer--Cope Thunder; Col Lee Hess--special operations; and Col Randy Miller--supply.

As the vice wing commander looked around the long conference table, it occurred to him that they were an exceptionally fine group of officers. Goodman had commanded three civil engineering squadrons before coming to Clark. His 3d Civil Engineering Squadron was the largest of its kind in the Air Force, numbering some 1,000 members.[24] Meeboer, Ray, and Hess were deeply experienced aviators with exceptional records. Duffy ran one of the largest hospitals in the western Pacific. In fact, every officer at the table had been handpicked for his job because of the large size of the units at Clark and the special demands of a large, remote overseas base.

The vice wing commander explained the charter from the wing commander: no loss of life; protection of property subordinate to keeping troops out of harm's way. They were there to formulate contingency plans for emergency response to the volcano. He reiterated. The first priority was no casualties. Not one. Nada. Zip.

Dave Harlow, from the USGS team, presented a brief summary of Pinatubo's activities. Rick Hoblitt, the rock hound, sat beside him. Harlow, who had no beard, had rotated in to replace Dr. Newhall as the team's chief. Hoblitt, Tom Murray, and Ed Wolfe were also relatively new arrivals. The USGS was planning for the long haul by changing out its people every 30 days or so. Harlow stressed the point that at any time all activity in the mountain could cease if the rising magma encountered a layer of rock that it could not fracture. On the other hand, if eruptions should occur there was no way at that time to predict when or how big they might be.

The volcanologists were using their standard warning system. The area around Pinatubo was at Alert Level 2, which meant the volcano could eventually erupt. Alert Condition 3, if implemented, would mean eruption was possible within two weeks. Condition 4 meant an eruption was possible within 24 hours, and level 1 meant an eruption was in progress.[25] The system was in use by both USGS and PHIVOLCS, and Harlow suggested the Air Force use the same system.[26]

Table 2-1
Clark Air Base Volcano Warning System

Concern	Moderate level of seismic activity Eruption possible within the next year
Caution	Volcanic unrest increasing Low possibility of eruption within two weeks
Warning	Volcanic unrest extremely high If seismic trend persists or increases, higher possibility of eruption within two weeks
Emergency	Intense volcanic unrest Eruption possible within 24 hours and evacuation determined necessary

The meeting consensus was not to use the same system. Base people were familiar with the typhoon warning system that numbered its levels exactly the opposite of the USGS-PHIVOLCS system. Ultimately, the base developed a simple scale to mirror the VolCon system.

The CAT discussed how to move people away from the eruption. Mount Pinatubo was 8.2 miles from base housing along the western wall. The real question was how far people would have to go to be safe.

Dave Harlow said it was hard to predict. Of course, everything depended on how big the eruption, if it erupted, would be. He apologized for seeming evasive, but there were many unknowns remaining, and even if a ton of data went through analysis, the best guess would come from past eruptions. Someone jokingly asked Harlow how far away he intended to be if it blew, and his laughing response was, "California!"

Rick Hoblitt's preliminary surveys showed that the newest pyroclastic deposits were in both riverbeds on either side of the base, but they did not go much past the very closest part of the base to Pinatubo. As for older pyroclastic flows, Clark was built on them.

Harlow indicated that 20 kilometers might provide a 99 percent certainty of safety. Immediately, several of the non-aviator CAT members suggested he speak in miles, not metrics. Harlow spread a large map of the area on the conference table and ran his finger in an arc 12 miles from the mountain. The 12-mile arc came through the flightline. He traced another arc at 15 miles. This line came through the base at nearly the east wall, area that was still on the base and the most distant from the volcano.

Everyone was standing, leaning over the large map on the table. The staff, several of whom were fighter pilots, were unwilling to accept a goal of 99 percent safety. One of them gave a brief tutorial on statistics. In the fighter business, pilots study their weapons in minute detail. Each can tell you, for example, that under certain circumstances, his heat-seeking Sidewinder missile has a probability of kill of 65 percent. One can view that number two ways. If you are the shooter, it means 65 percent of the time you shoot down the target. If you are the other pilot, though, in the target airplane, it means that 65 percent of the time you are 100 percent dead.

In a perfect world the evacuation would occur over a weekend. Most thought it was far too much to hope for

Everyone agreed that any plan must move people to an area where they would be safe; not statistically safe, perfectly safe. For the time being, the 12-mile arc, which was in a very large open area at the extreme eastern edge of the base looked good.

A long discussion ensued about moving people to safe areas. At this point, all the discussions were centered on a scenario in which the mountain exploded in a huge blast as had Mount Saint Helens. Everyone, including the volcanologists, viewed that as the worst possible case.

A warning system was already available. The base was equipped with a siren system mounted on towers throughout the facility. Whenever the base exercised its war time ability to survive and operate while under enemy attack, the sirens sounded alerting all of an imminent air attack. Also, there was a "Giant Voice" system of loudspeakers mounted on the same towers that could be used to pass instructions. Giant Voice was seldom used, but it was operational and available. Of course, FEN television and radio was available in every home and work area. Therefore, warning people to move to a safe area was no problem--if there was enough time.

The scientists and the colonels discussed the time factor and the warning system. The military men, in their continuous quest for cut-and-dried answers, wanted the USGS scientists to be specific about how they thought things might progress.

The beards resisted. They encouraged the military men to slow their thinking a little. The issue was a volcano. Geological events usually proceed slowly. The process of moving to an eruption could take many, many months if it happened at all. Although the activity at Pinatubo clearly was moving at a relatively rapid pace, the pace could slow, or stop, or even reverse itself at any time.

A very important point was made; one that would play throughout the Pinatubo experience. Those who did not understand this point would never understand some important later decisions. Harlow explained that if an alert level was reached that indicated the possibility of an eruption within 24 hours, but the volcano did not erupt, then there could be a very long period before the USGS would be confident enough to back the level down to a lower one. There was no way to predict how long that would take. The USGS warning system indicated the delay might be as long as a week, but the scientists were skeptical it could be done that quickly.

The implications of this point were clear to the CAT. If the mountain got to what the Air Force had termed the "Emergency level," and the CAT ordered everyone to go to an area across the runways, then they could be there for weeks.

The CAT discussed the possibility of a tent city but quickly dismissed the idea as impractical. A tent city was sustainable for the short term but not for weeks on end. An evacuation for a short time to a tent city might be reasonable, but it seemed to the CAT that a second evacuation to a more sustainable location must follow. To where? How?

The "how" was easy because there was only one answer. Each person would have to drive his or her privately owned vehicle (POV). It was the only way to move 15,000 people quickly. Moreover, people could carry whatever equipment and supplies they needed and wanted in their cars with them. Children would be with parents. For those who did not own cars, mostly young airmen who lived in the dormitories, there were enough motor pool buses to handle them.

The "where" question was much more difficult. One-by-one the staff eliminated possibilities. Wallace Air Station, a small USAF facility a few hours drive to the north was far too small to handle the numbers of people involved.

Manila was only two hours drive away, but the prospect of dumping 15,000 people there was too complex to organize under the best of circumstances, let alone in an emergency. The services squadron had already checked with hotels in Manila. They would not make room by canceling other customer's reservations even if the base was in an emergency situation.

Finally, the CAT staff agreed that Colonels Murphy and White had chosen the logical alternative: the Navy complex at Subic Bay. Subic Bay was about a two-hour drive and some 25 miles from the volcano. Two bases sat on the bay: Naval Station Subic Bay and Naval Air Station Cubi Point. Even though the bases were not large in area, they held about the same population as Clark. At least everyone could have shelter there until the situation stabilized at Clark. Once the smoke cleared, several options would be available.

If Clark was habitable, a re-occupation of the base was possible by slowly bringing people home. However, Dave Harlow's point was well taken that it could be a long stay until USGS backed off its warning level, especially if an evacuation to Subic Bay was ordered and the mountain did not erupt. Since Naval Station Subic Bay was a deep-water port and the adjacent Naval Air Station Cubi Point had an airstrip, a permanent evacuation to the United States could be staged from Subic Bay if Clark was uninhabitable, or if the emergency level held steady for an extended period.

The group continued to discuss the myriad problems to be faced. Col Brian Duffy, the hospital commander, expressed his immediate concern for the non-ambulatory hospital patients. Clark was home to a huge regional medical center. They would have to be moved early since they could not respond quickly.

Duffy did some quick math aloud concerning pregnant women. From a population of several thousand young families, how many women are in their last tri-mesters, or even their last month? Answer: dozens. Duffy added that he had already started the wheels rolling to earmark medical evacuation (Medevac) aircraft to move patients on short notice.

Colonel Murphy, the support group commander, or city manager as civilians would call him, was an outstanding officer. Young, energetic, and full of ideas, he enjoyed an excellent reputation throughout the base population. He always seemed to know what people were saying and thinking in the community, and he was very bothered about the lack of information on the volcano. Murphy outlined the spadework that he and White had already completed. He suggested that the leadership start making public service announcements on FEN about what was happening with the volcano.

The CAT commander rejected Murphy's suggestion. Information about the volcano was already being passed through the chain of command and other means such as commander's call, staff meetings, first sergeants' meetings and chiefs' meetings. Murphy did not think this method was working. He stated that rumors were flying. People were stopping him on the street to ask what was going on with the volcano. Murphy's position was that the rumors were making the volcano worse than it actually was.

The volcano was at the bottom of the warning hierarchy for both the USGS and the base warning systems. The scientists were still gathering information and were uncertain which direction it would go. There was a fine line between preparing for a possible evacuation and throwing everyone into a "get me on that lifeboat" state of mind.

The task of controlling traffic fell to Col Bill Dassler, Clark's security police group commander and "Top Cop." Dassler was a superb leader and commander. He was a very big man with a full shock of white hair, bushy eyebrows, and confidence in his ability to lead his men. He smoked unfiltered Camels in a day when it was politically incorrect to smoke. Someone long ago had told him the secret to leadership was to keep your men out of the sun and make sure the horses have water. He believed it and practiced it every day. When he said in his gravelly voice that he could get something done, all believed him.

His first job was to conceive a plan to get everyone to the flightline area that was just outside the 12-mile ring from the volcano. The aircraft-parking apron provided a vast expanse of concrete to marshal vehicles for further movement to another area. Dassler was very confident at the first planning meeting that his security police could handle any traffic situation they might encounter. Some were concerned that once the evacuation started the 250,000 locals in towns surrounding the base would panic and take to the roads at the same time.

"Don't worry about that," said Dassler, "we're already working with the INP (Integrated National Police) to make sure the roads stay open."

After much discussion, the CAT agreed upon a philosophy that would drive future planning. If the mountain progressed to the point that dictated an evacuation, it was all or nothing. Once people started moving, they would have to keep moving to a place where they could be sheltered and fed for an extended time: Subic Bay.

Murphy stopped the meeting cold for a moment when he asked, "What if we have to do this during the day?" Husbands would be at work, children in school, people in the base exchange and commissary, mothers out running errands, house-girls minding small children, off-duty troops playing golf. It would be "a chocolate mess," as the world famous fighter pilot, Steep Turner, would say.

The group determined that an evacuation, under the best circumstances, must be announced the night before then executed early in the morning while most families were together. They could pack their cars and leave together. If they had to evacuate suddenly, during a workday, it would be chaotic with mothers trying to find children, husbands trying to find families. In a perfect world, the evacuation would occur over a weekend. Most thought it was far too much to hope for.

Bricks

Col Al Garcia, communications squadron commander, reviewed the emergency communications capability of the base. A very capable hand-held radio network of 600 "bricks," called a trunked land mobile radio system, was already in place. "Brick" was a nickname that had originated in the day when hand-held radios were the same size and weight as a brick. Clark's were a new style that gave users the capability not only to talk with each other in walkie-talkie fashion, but also to make telephone calls.

Brick users used call signs rather than names, ranks or duty titles. At most Air Force bases, people on the command net used a common call sign followed by a number which indicated the user's place in the pecking order of rank structure. For example, at Seymour Johnson Air Force Base, North Carolina, the wing patch featured a fierce lion. So, the wing commander's call sign on the command net brick was Lion 1; the vice wing commander was Lion 2, and so on. At Clark, security concerns dictated that call signs be less obvious so that NPA terrorists would not be able to intercept a radio call and know who was making it.

At Clark, General Studer was Gator; Colonel Grime was Eagle. The CAT commander (Colonel Anderegg) was Lucky, a nickname he was given after narrowly surviving a crash many years earlier. Dassler was Gladiator. Those working in the command post used Hawkeye. The bricks that Col Al Garcia, Wizard, maintained saved many lives and became nearly as important as potable water. As the crisis unfolded the command structure of Clark came to depend totally on the brick network, and as the network expanded everyone on the base came to know General Studer as Gator. Of course, military courtesy and discipline prevented anyone from daring to call him that personally, but when Gator, Eagle, or Lucky spoke on the net, everyone knew exactly who it was.

Strolling

Even while Clark's leaders agonized over scenario after scenario, the daily routines continued for the general population. High school seniors looked forward to their graduation from Wagner High School. Filipino house girls played with children in playgrounds and yards across the sprawling base housing areas. In the evening, whole families strolled under the giant acacias around the parade field. Some stopped for an ice cream cone. When they stopped to chat with others, the first question invariably was, "What have you heard about that volcano?" It was the question on everyone's mind. As the plume of steam got larger, so increased the uncertain, uneasy feelings of everyone who looked at the ominous beacon every morning on the way to work, school, or play.

NOTES TO CHAPTER 2

1 Interview, James Boyd with author, November 2, 1998, notes on file AF/HO, Bolling AFB, DC [hereafter cited as Boyd interview].

2 Interview, John E. Murphy with author, November 10, 1998, audio tape at AF/HO, Bolling AFB, DC [hereafter cited as Murphy interview].

3 Interview, Dr. Raymundo S. Punongbayan with author, November 2, 1998, audio tape at AF/HO, Bolling AFB, DC [hereafter cited as Punongbayan interview].

4 Punongbayan interview.

5 E-mail, Christopher G. Newhall to author, October 21, 1998 [hereafter cited as Newhall email].

6 Ibid.

7 Author's journal.

8 Fire and Mud, p 215.

9 Newhall email.

10 Interview, Major General Jeffrey R. Grime with author, February 22, 1999 [hereafter cited as Grime interview 1999].

11 Journal, Richard P. Hoblitt, David A. Johnston Cascades Volcano Observatory, 5400 MacArthur Boulevard, Vancouver, WA, 98661 [hereafter cited as Hoblitt notes].

12 Fire and Mud, p 210.

13 Author's journal.

14 Letter, Dr. Christopher G. Newhall to C. R. Anderegg, 27 July 1996 [hereafter cited as Newhall letter, 27 July].

15 Interview, Col Jeffrey R. Grime with Dr. Tim Keck, June 10, 1992, transcript and audio tape at PACAF/HO, Hickam AFB, HI [hereafter cited as Grime interview 1992].

16 Newhall letter, 27 July.

17 Ibid.

18 Hoblitt notes.

19 Murphy interview.

20 Ibid.

21 Grime interview 1992.

22 Ibid.

23 Ibid.

24 Goodman interview.

25 Fire and Mud, p 90.

26 Philippine Flyer, Vol. 39 No. 22, 7 June 1991, p 5.

Chapter 3 DRUM WATCH

Throughout the rest of May, commanders and scientists watched the volcano very closely from the base house, named PVO, where the volcanologists lived and worked. On May 17, the USGS-PHIVOLCS team met with senior officers on the base to review the situation and discuss possible scenarios. By this time, the scientists had confirmed what they suspected earlier in the month. Pinatubo displayed every sign of being an explosive volcano. They were unsure if the volcano would erupt, but if it did, it would be a big one.[1]

They presented a probability tree – a schematic drawing that displayed the possible courses the volcano might take and the probabilities of each course. The discussion about the probability tree was academic, but the ensuing information given by the geologists emphasized the gravity of the situation.[2] They confirmed that pyroclastic flows had come down the Clark side of the volcano. All the prehistoric eruptions were big. There was a chance, albeit small, that the base would be swept by pyroclastic flows.

Within the base's general population, few were able to grasp the concept that what looked like a small hill barely discernable from its neighboring mounds was a real threat to life and property. Additionally, the farther one got from Clark the harder it seemed to comprehend a scenario that might require an evacuation of non-essential military personnel and all dependents.

In mid May, General Jimmie V. Adams, the Pacific Air Forces Commander, visited Clark Air Base. It was his first visit as the PACAF Commander to Clark, and it was part of a routine first visit to his bases in the Western Pacific. Part of his itinerary called for a helicopter tour of the Cope Thunder exercise area along Crow Valley. He was escorted on the flight by General Studer. As they left Crow Valley, Studer told Adams he wanted to show him the steam coming out of Pinatubo, so the helicopter flew the few miles to the volcano. General Adams saw the steam and could smell the sulfur in the air. He asked Studer what it meant. Studer told him, "Well, they think this is going to be an active volcano soon." When General Adams then asked how long it had been since the last eruption, the Thirteenth Air Force Commander told him, "600 years." General Adams recalls not being concerned at that point.[3] Following General Adams' visit to Clark, the CAT commander received numerous calls from officers stationed at the PACAF headquarters in Hawaii. Many were curiosity seekers, some offered assistance, but a few carried a common message. "Chicken Little said the sky was falling, and it didn't. Are you that worried?"[4]

Geology 102

Geology lessons from the volcanologists continued. They discussed ash fall, which usually was like talcum powder with a very fine texture and very abrasive. If a car engine ingested it through the air filter, the abrasion scoured the cylinders and ruined the engine. The base was already receiving documentation from Air Force bases in Washington State and Alaska that had experienced a light dusting of ash from nearby volcanoes. People near Mount Saint Helens found they needed to change their oil and air filters every few hundred miles.

The eruption became more evident...

Because the ash was so fine, it would work its way into everything, including computers and electrical equipment. It was highly conductive, so it could easily short out electrical equipment. Telephone systems and electrical transformers were especially vulnerable, the beards lectured.

The ash itself was not life threatening, although people in the area might experience upper respiratory irritation and eye irritation. They added that inhaling the ash might exacerbate asthma or other chronic breathing problems.

Of course, the most dangerous aspect of an eruption was the pyroclastic flows that can come boiling down the mountainside. However, the lesson continued, another major danger was rain. If the ash is wet, it could be very dangerous because it could hold a lot of water. Newhall was very concerned that wet ash might cause real problems in the local community since so many of the buildings had flat roofs. Wet ash might cause roofs to cave in if there is a lot of it. Wet ash was like very wet beach sand, holding a volume of water nearly equal to its dry volume.

...some thought we were over-reacting and some thought we were under-reacting

One scientist offered an example. Place two empty drinking glasses on the table. Fill one with dry, powdery sand and the other with water. Now dump the water from one into the other with the sand. The sand will absorb all but a small amount of the water, and the weight of the glass will be equal to both before the water was transferred. Now, imagine a few thousand of those glasses standing on a roof.[5]

Newhall brought a videotape that demonstrated volcanic hazards. The tape showed enormous pyroclastic flows of gray, superheated clouds of ash spilling down mountainsides. Clearly, nothing could survive such an onslaught. The camera panned across wide fields of pyroclastic deposits left in the wake of the pyroclastic flows. Everything was buried and the flow was still smoking.

If the views of the pyroclastic flows were not sobering enough, the tape went on to discuss the dangers of lahars. If ash on the ground gets wet enough, the whole mess will run through watersheds like flash floods of wet concrete. The phenomenon was called a lahar. If the ash was still hot, it was a hot lahar.

During an eruption, the rising heat from a volcano frequently produced large thunderstorms near the volcano. As the ash covered the ground, rain from the thunderstorms would supersaturate the ash. If the whole mess became heavy enough and wet enough, it would start running down rivers and streams with sometimes devastating results.

Lahars could be a deadly problem for years, perhaps decades, after a volcano stopped erupting. Large ash and pyroclastic flow deposits could absorb vast quantities of rainwater. If the ash became wet enough, huge slabs, like icebergs of wet cement, might scab away causing an instant flash flood. Nonetheless, everyone's main concern was the devastation pyroclastic flows might cause. Since it was the dry season, lahars were a secondary concern.[6]

The video, which had everyone's undivided attention, was made by Maurice Krafft and his wife. The Kraffts were famous in volcanology circles for their spectacular photography of erupting volcanoes. It was easy to see from the video that the Kraffts had put themselves in some very dangerous spots over the years to record the breath-taking images on the tape. The Kraffts were scheduled to come to visit their friend, Chris Newhall, at Clark late in May. They never made it. They were killed in Japan by a pyroclastic flow from Mount Unzen. They died at nearly the same time their video was educating the people of Clark Air Base and the surrounding Filipino communities.[7]

Colonel Grime ordered a comprehensive education of the base to begin. Colonels Murphy and Rand started an all-out program. They took briefings to every organization on and off the base they could schedule. They went to first sergeants' meetings, school teachers' sessions, the Retirees' Activities Office, commanders' calls and recreational club meetings. At every meeting, they encountered deep skepticism. Murphy had a series of slides; taken at five-day intervals that showed how the steam plumes had grown and become darker as ash mixed with the steam. Some listeners took him seriously, but many crossed their arms and appeared unconvinced. The community was callous to emergencies, or in this case, possible emergencies.[8] The year-long series of earthquakes, typhoons, government worker strikes, assassinations, and frequent restrictions to the base and quarters had taken its toll on morale. It was very low. Murphy, Rand, and others who tried to educate people about the volcano often met blank stares that said, "Not again!"[9]

Those who listened with care and open minds realized the danger, and what they heard scared them.[10] Rumor control was a primary function of the public affairs office that was led by Colonel Rand. One tool they used was a rumor control hotline that concerned people could call and leave questions. The base newspaper, the *Philippine Flyer*, then published a representative cross section of answers in an attempt to disseminate the truth. The answers were given in a section of the paper named "Just the Facts." A longer version of questions and answers was broadcast on an FEN radio program that Rand personally hosted. In a very short time, the hotline at "Just the Facts" was saturated with questions solely about the volcano.[11] At its peak, the hotline was getting 600 calls a week, an "astronomical" number.[12]

Despite these efforts to get the ungarbled truth in front of the base populace, the troops did not believe they were getting enough information.[13] By the end of May, the volcano was belching small clouds of ash that everyone could see, and the volcano became the topic of conversation at every opportunity. Colonel Grime was off the base on a business trip, and Colonel Murphy approached Colonel Anderegg, the vice wing commander, about putting a USGS scientist on FEN-TV to educate those who were not getting the word through the chain of command. Anderegg agreed.[14]

Colonel Rand's own words best describe the situation. "At the end of May. . . everything we were doing didn't seem to be providing enough information . . . they wanted more . . . because the mountain was beginning to look more volatile. We decided to have a news special, an evening news special in which I went on as the host and head geologist Chris Newhall went on with me, and we spent the first 25 minutes talking about what's going on with the mountain and what it might be doing."

"We played it on a Thursday evening," Colonel Rand continued, "and it was very well received. People called in and asked that it be replayed. So I did once a day or once every other day, people called in, put it on at different hours . . . personal contact with groups like churches, wives clubs, wherever people were gathered together in groups. I can remember schools on base, the elementary, middle and high school asked us whether we could come out and talk. We did that. Anybody wanted somebody to come out and talk to them, we did that. We worked the newspaper hard; we worked the chain of command . . . the first sergeants and chiefs. General Studer's philosophy was you ought to have one spokesman and that way one message gets out so basically when it was time to talk about the volcano or the plans, he (Studer) would say, okay, and we'd tell them what we know."

In this venture Colonel Rand had unlimited access to General Studer, Colonel Grime, and Colonel Murphy. Colonel Murphy actually hosted the first sergeants' meetings. Colonel Grime, the wing commander, had the commanders' meetings. They would each make their pitch and make their points and make their message, but then they'd make sure that the public affairs spokesperson had an opportunity to say that's exactly what we told the chiefs.

Colonel Rand concluded his assessment in this way: "So, we worked it pretty hard, so I think we had a pretty informed community . . . yet a wound-up community and the winding-up was, I think, . . .kind of the icing on the cake and that really put a lot of people over [the edge]. The horrible anxiety. But you never had a consensus. With the restrictions or with the volcano . . . some thought we were over-reacting and some thought we were under-reacting."[15]

Newhall's assessment of the effect of his televised presentation differs somewhat from Rand's. Newhall wrote: "My interview on FEN around the end of May seems, in retrospect, to be too reassuring . . . . I could and should have been more blunt about the possible risk, but I don't think I could have been any more definite about what was to happen, or when. At that time, the volcano hadn't started down its unequivocal path to eruption, and we try hard to avoid false alarms because the next time, when people really need to pay attention, they won't."[16]

Credibility was a great concern for both the USGS staff and commanders at Clark. Newhall continued, "The odds were against a successful forecast. And, even if we got the forecast right, we still had to worry about whether people

The Silver Wings Community Recreation Center.

would actually heed the warnings. From my perspective, it was a very close call. You'll recall the skepticism on the base, and that was just as high among politicians and most other community leaders outside the base. The Krafft video slowly began to grab people's attention, as did a variety of other little things on and off base . . . we had the huge advantage for folks at Clark that, if the command said move, everyone would move. No endless debate."[17]

Grime, the wing commander, felt the same pressure. If he ordered an evacuation, and the mountain did not erupt, all credibility would be lost. "I had this vision," Grime said, "of one night sitting there in bed and . . . seeing 15,000 people in front of the house with tar, feathers and a pole escorting me off the base after they had returned."[18]

> *...Colonel Grime continued to be concerned that an evacuation under panic conditions could kill more people than the volcano*

General Studer called his boss, General Adams, and told Adams that the base was making plans to evacuate. General Adams recalled, "I'm thinking he is a good conscientious commander, but maybe he is leaning a little too forward in the chocks. I don't understand what the real problem is. My frame of reference, having been down to the big island [where] you see the molasses coming down the volcano when it finally erupts. And it takes days and days for it to finally get down to the ocean."[19] Studer told General Adams that Dr. Newhall, who was rotating back to the states, would stop in Hawaii and brief him. The briefing had the desired effect. Adams remembers that Newhall told him the base could be engulfed in pyroclastic flow within 20 minutes of an eruption.

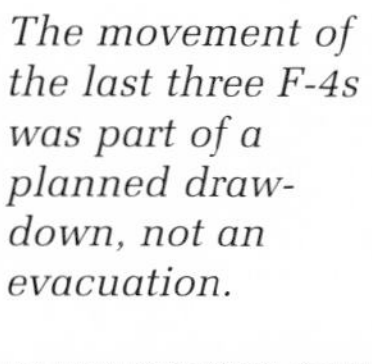

The movement of the last three F-4s was part of a planned drawdown, not an evacuation.

Last of the Phantom Jets

By the end of May, an eruption still did not appear to be imminent, and the geologists could say little or nothing about how much longer the unrest could continue before escalating or dying off. However, over the next several days, the earthquakes nudged even closer to the surface and the amount of ash in the steam plume gradually increased.[20]

By June 4, 1991 only three of the 3 TFW's F-4, Phantom II, fighter jets remained at Clark. The others had not been flown out to avoid the volcano; rather, the elimination of the aircraft was part of a planned drawdown directed by the Pentagon the previous year.

F-4 Phantoms were the backbone of PACAF's fighter forces since the end of the Vietnam War, and the ones at Clark were PACAF's last. Throughout the first months of 1991, the proud war birds flew by twos and threes back to Davis Monthan Air Force Base, Arizona, there to be decommissioned and interred in the bone yard for excess Air Force equipment.

The flight out of Clark that morning was led by the 3 TFW vice wing commander, Col Dick Anderegg, and his back-seater, Maj Hugh Riley, in F-4E tail number 1085. Maintenance crews had buffed the jet to a high sheen, and as Anderegg taxied slowly out of the aircraft's protective shelter, they were greeted by waving American flags and waving people along one side of the taxiway. On the other side of the taxiway, maintenance troops stood at rigid attention, shoulder to shoulder. Each snapped off sharp hand salutes as the flight leader and the other two jets moved slowly through the cordon of well-wishers.

The flight leader's call sign was Clan 11. In Clan 12 were Capt Mike Trinchitella and his backseater, Capt Robert Ricarte; and Clan 13 was flown by Maj Howard Hendricks and his backseater, Maj Jimmy Miyamoto. The jets thundered down the new runway one by one, then as the leader made a gentle left turn to the north, the wingmen joined up in close formation. After another left turn to align with the runway, Clan 11 flight flew by the base and departed to the southeast. There was not a dry eye in the house.[20]

Hurried Preparations

Meanwhile, evacuation planning was going on at a frantic pace. Colonels Murphy and White, no longer operating covertly, worked with Col Brian Duffy, the hospital administrator. On the same day the F-4s left, several thousand patient records from the hospital were gathered up, and loaded onto buses. All were airlifted out of Clark to other Pacific bases for safekeeping.[22] Duffy, like Murphy, was being as pro-active as possible.

On June 5, increasing numbers of earthquakes, an ominous decrease in the amount of sulfur in the air, and increasing ash accumulations at Pinatubo's summit spurred Dave Harlow to tell Colonel Grime that the team was upping the alert level to 3, indicating a 50 percent possibility of an eruption within two weeks. Part of this action was based on the observation of a "spine" on the volcano's flank that could have been an indication of magma just below the surface. John Ewert's tiltmeters and other indications showed the mountain was bulging a little. However, the warning also acknowledged the possibility that they might have to go back down to level 2. The next day, Rick Hoblitt looked more closely at the "spine," first noticed by Andy Lockhart, and realized it was an old rock formation, a remnant. However, even though the team lost a little credibility with the wing commander, Colonel Grime, they elected to stay at alert level 3 until other indicators had an opportunity to show that the mountain was cooling down. That never happened. In fact, the seismic activity indicators continued to increase, and the sulfur emissions decreased dramatically – a possible sign that the mountain was building pressure rather than venting it into the air.[23] About 10,000 Aeta tribesmen moved from their homes high on Mount Pinatubo into the evacuation camps located in safer areas.[24]

Clark commanders were caught flat-footed by the sudden increase in volcanic activity. Two days later, in the last regular edition of the *Philippine Flyer*, the base warning level was still shown at the benign "Concern" level, which indicated an eruption possible within the next year.[25] However, the paper also contained a volcano checklist and a checklist for departing housing.*

The volcano checklist was merely a modification of existing checklists contained in worldwide NEO (non-combatant evacuation order) plans designed to evacuate dependents from zones threatened by combat. The checklists were also sprinkled with advice from pamphlets Colonels Murphy and White obtained from Air Force bases that had experienced some light ash fall in the United States. For example, operating air conditioners would suck the ash indoors. If the air handlers pulled enough ash inside, then electric appliances might short out. Instructions to leave some interior lights and all exterior lights on were there to assist security forces' efforts to control looting.

* *Tables 3-1 and 3-2 contain these checklists.*

Table 3-1 Volcano Checklist

Cash for each family member (both cash and pesos)	Inventory of all household goods
ID cards, passports/visas	Wills
Ration cards (CEX)	Personal medications
Immunization records	Infant care items (diapers, toys, formula, portable crib, stroller, etc)
Birth certificates	Water supply for one day
Marriage certificates	Food/snacks for one day
Toiletries	Food for pets (3 days) and leash
Three sets of clothing for each family member	POVs should have one-half to three-fourths tank of fuel
For pregnant women: physician's certification of pregnancy, stating expected date of delivery	Flashlight with spare batteries
Pillow and blanket or sleeping bag for each individual	Portable radio with spare batteries
Vehicle registration and car insurance policies	Candles/matches
Checkbook/bank book/credit cards	Bug repellant
Personal insurance policies	Sunscreen, hats, etc.
	First aid kit, knife, rope, tool kit
	Toilet paper

Alongside the checklist, words from the public affairs office described the situation. "Scientists monitoring the mountain recorded a slight increase in seismic activity recently. However, based on information provided by the experts, base officials see no requirement to change Clark's alert level at this time. Consequently, we're still at the concerned level, which is a planning stage. The experts remain confident that we're not in imminent danger and that we'll have plenty of advance warning before a large-scale eruption."[26]

Suddenly, a conflict between the USGS warning system and the preparations of the base became evident. The USGS warning was at least two steps ahead of the one that base commanders were using. However, the base was not prepared for an evacuation. No real plan had been revealed to the base populace, and Subic Bay was not prepared to accept 15,000 visitors in one gulp.

Table 3-2 Steps for Leaving Quarters

* Turn off all air conditioners/central air or other ventilation devices.	* Close all windows.
* Turn off electric ranges, washers, and dryers. Keep refrigerator and freezers on.	* Leave some interior and all exterior lights on.
	* Lock all external doors.

Colonel Murphy, Colonel White, Maj Shevon Peete, Lt Col Ed Connor, and others doubled their efforts to prepare for an evacuation. Peete and Connor returned to Subic Bay to discuss the situation with their counterparts. Murphy and White coordinated with Colonel Dassler, the "Top Cop," their ideas of how best to move the largest Air Force overseas installation out of harm's way at a moment's notice.

They divided the base into sectors and prioritized each according to its proximity to the volcano or the likelihood of pyroclastic flow or lahar exposure. The plan would be to get Mactan housing, which was only 8 miles from the volcano, out first. The Hill residents would be followed by housing areas on the base's flanks that were closest to the rivers on either side of the base.

Each area would have 90 minutes to evacuate before the following area hit the roads. Each family would drive its own car and carry anything they wanted as long as it fit in the car. The evacuation route would flow down three main streets of the base then out onto the aircraft parking apron. By this time, all the 3d Tactical Fighter Wing aircraft were gone, and the 353d Special Operations Wing was moving its aircraft with the last scheduled to depart on June 9. There was plenty of room to marshal the vehicles on the flightline and then move them to a choice of gates exiting the base. The choice of gates would be made based on what the volcano was doing. If it were erupting, gates near the river would be closed, although some of them were a more direct route to Subic Bay.

While cars were marshaling on the parking apron, workers would gather data from each car such as the occupants' names and relationships. Water and other simple, but valuable, supplies would be issued to each vehicle. Evacuation planners had no doubt traffic on the highway to Subic would be very slow. The road was in poor condition, and the traffic on it was always slow. Another few thousand cars entering the highway likely would bring the pace to a creep.

Dassler and his men coordinated with the Filipino Integrated National Police and the Armed Forces of the Philippines (AFP) police to control the traffic. Most planners expected thousands of locals would follow the Americans to the Navy base southwest of Clark thus further congesting the clogged highway.

The confusion between the USGS warning system and Clark's continued. On June 7, a dome appeared on Pinatubo's northwest face. Increased seismic activity and the extruding dome prompted the volcanologists to increase the warning level to alert level 4, possible eruption within 24 hours. Clark commanders increased their warning level to "eruption possible within 2 weeks."[27]

View of the gate at Clark AB as personnel were evacuated before the major eruptions and before the ash fell.

All of this is not to say Clark commanders were overriding USGS scientists' decisions. Rather, the two had different agendas. The USGS-PHIVOLCS team was concerned about Clark's people, but they were equally concerned about the large Filipino population that was much closer to the volcano. Even a small eruption could kill many thousand locals. The volcano, even as late as June 7, did not show signs of a massive eruption that would threaten Clark. Therefore, Clark officials, who were concerned only for the welfare of its charges, were trying to determine one thing and one thing only: when to evacuate.

The political implications of an American evacuation were enormous. The continuing discussions over basing rights and the cost of those basing rights were tenuous. Many believed the Philippines Senate would not vote to ratify a new agreement with the United States to keep Clark and the Navy bases at Subic Bay open. A premature evacuation of the base might be interpreted as a signal that the Americans were trying to bully the Philippine Senate by showing them how bad the economy might become without Americans spending their dollars there.

General Studer, who felt no political pressure, was not going to order an evacuation until he was sure the volcano might become a threat to American lives.[28] For his part, Colonel Grime continued to be concerned that an evacuation under panic conditions could kill more people than the volcano.[29] Therefore, the disagreement amounted to definitions of the size of the threat. Any threat near the volcano triggered a USGS elevation of the threat level. Clark, located eight miles from the volcano, seemed to have a larger margin of safety.

By June 9, the margin narrowed considerably. Pinatubo was getting hotter by the hour. Earthquakes were coming in swarms across the drums at PVO. A pyroclastic flow was reported

US Air Force Hospital at Clark AB.

in the Zambales, away from the base. Technically, this amounted to an eruption, albeit a very small one, and the USGS-PHIVOLCS team upped their warning level to 5, eruption in progress. General Studer and his forces stood pat.[30]

Planning at Subic Bay was going at a frantic pace. However, everyone there representing the Air Force knew they were woefully behind. The fast pace of instability in the mountain had caught up with them. Subic Bay was already short 800 beds for the forces stationed there. The two Navy bases' electrical system was stressed to the maximum, and large sections of the power grid had to be turned off everyday in order to give all parts of the bases a "fair share" of power. Water was rationed as well.[31] The thought of adding 15,000 more was nearly incomprehensible. Plans were made to use facilities at Naval Air Station Cubi Point, the Naval airfield adjunct of Subic Bay, and housing in the San Miguel housing area a few miles north of Subic Bay, also owned by the Navy.[32]

Rand looked down at the piece of paper Studer had handed him. It was the evacuation order

Colonel Murphy and his group produced a pamphlet that described the base evacuation plan. Between June 7 and June 9 many thousand were produced. Couriers delivered a pamphlet to every home on the base and all the ones off base that could be located. They placed stacks of the pamphlets in the base exchange, the commissary, all the recreational clubs, the chapels and work centers throughout the base.[33]

For several days, commanders on the base were determining which people should remain on the base. After several iterations, Colonel Grime approved a listing that included Dassler's entire security police group of 960; a contingent of civil engineers with expertise in electrical power production, heavy equipment operators, and fire department personnel; cooks; communications specialists; and the CAT, minus its commander who was rushing back to Clark after delivering the wing's last F-4 to the bone yard. Altogether, the forces that were designated as "mission essential" numbered about 1,200. The mission essential team would not evacuate.

On the morning of June 9, General Studer directed Colonel Rand to inform the base that a decision would be made early the next morning concerning the need to evacuate. The base caught its first lucky break. June 9 was a Sunday. The vast majority of the base population was in or near their homes.[34] Rand made an announcement over FEN television and radio that told everyone of the situation. In his bland monotone, he said further announcements would come that evening, after the scientists and base commanders had the opportunity to study the volcano further. By 7:00 that evening nothing had changed with the volcano. Rand told everyone to continue preparations for an evacuation and another announcement would be made at 10:00 p.m.

Shortly before 10:00 p.m., General Studer delayed his evacuation decision until early morning on June 10, the next day. Consultations with the USGS indicated that there was no basic change in the condition of the mountain. Colonel Rand prepared to go on the air and pass along that information, but shortly before airtime, the television studio lost electrical power. It took nearly an hour to fix the problem; and by the time Rand transmitted the news, he assumed he was sending it to a very hostile audience. He said that the decision would be made early the next morning and in time for the first evacuees to depart their homes at 6:00 a.m. He encouraged everyone to double check that every family and each individual had the items ready that were on the evacuation checklist.

Colonel Rand was upset that his credibility had suffered due to the long delay before going on the air, and he was concerned the power failure might re-occur right when the evacuation order was given. He ordered the station crews to stay on duty all night and make sure the studio was ready to go at a moment's notice.[35]

Hitting the Road

General Studer met with Colonel Grime, Colonel Rand, the USGS team and others at 4:30 a.m., June 10. Studer asked questions of the volcanologists for about 20 minutes then turned to Rand and asked him for the two announcements Rand had prepared. One announcement said there was no change in the volcano, so stand by. The other said the evacuation would begin at 6:00 a.m. Studer looked at the two announcements for a moment then handed one to Rand. "Go with this one," the general said. Rand looked down at the piece of paper Studer had handed him. It was the evacuation order.[36]

At 5:00 a.m., June 10, 1991, Lt Col Ron Rand, "Colonel Bland," came on the air and started his broadcast the way he had started other broadcasts, "Good morning, Clark!" He reiterated the evacuation instructions the way he had given them the night before and the way they were written in the evacuation pamphlet.

The first evacuees left their homes at 6:00 a.m. Immediately the streets were flooded with cars, but Colonel Dassler's cops kept them moving toward the flight line and the marshaling area.

On the flightline, Murphy's services troops and morale, welfare and recreation (MWR) people passed out bottled water, snacks and other comfort items. The lines on the parking apron were long but kept moving. Personnel specialists and security policemen gathered information from drivers that indicated who was travelling with them. Base buses traveled through the dormitory areas and picked up single airmen who did not own cars.

Most cars were heavily loaded. Besides family members and pets, many Americans brought their domestic help as far as the gate. There, many said a tearful goodbye to people who had become de facto family members. Some did not drop off the domestics but continued to Subic Bay with them rather than abandon them to an uncertain future.

No one was happy about leaving, and the skeptics were not shy about saying so. Eagle, the wing commander, recalled, "When they were all going out the gate, it was a beautiful day. A bunch of [young airmen] stopped at the gate, and one of the [news] correspondents stuck a microphone in the window and asked what they thought of [the evacuation]. They just laughed and said, "Look at what we've got today. This is our standard leadership, do you believe they are having us do this?'"[37]

By most accounts, the base was empty by noon. Nearly 15,000 people had left in six hours. The highway to Subic Bay was bumper-to-bumper with American cars. The leading edge of the onslaught hit Subic Bay at about 8:00 a.m. Whatever plans Subic Bay had soon foundered under the relentless waves of Clark refugees streaming onto the small bases.

Hi! I'm from Clark. What's for Breakfast?

Col Jim Goodman, commander of Clark's civil engineering squadron, had traveled to Subic Bay the night before, so he was able to watch the evacuees pouring onto the base. Navy officials had designated the enlisted sailors' club, the Sampaguita Club, as the central reception point for the evacuees. Goodman watched in amazement as lines of people in front of the Sampaguita Club grew to nearly half-a-mile long.[38] Inside the club Navy and Air Force personnel and billeting specialists labored mightily to find shelter for the evacuees. They assigned several hundred families to unoccupied homes in the San Miguel housing area.[39] Several hundred were sent to the Navy recreation area on Grande Island. They sent others to the few open billeting rooms available since quarters at the Navy bases were already 99.5 percent occupied.[40] Navy families from Subic started taking evacuee families home with them. These efforts did not make a dent in the ever-lengthening lines that went out the door, around the corner and several blocks down the street.

In an attempt to shorten the lines and speed up the process, hundreds of evacuees were directed to the base's football field. They were instructed to wait there while one family member went to wait in the lines at the Sampaguita Club. Those who waited and sat in the field's bleachers sweltered in the summer sun for hours. Neither water nor shade was available. Some sought respite by crawling into holes they dug in the sandy soil under their cars.[41]

Other issues besides water and shade arose. One of the biggest was the pets evacuees had brought with them. Wartime NEO plans called for pets to be euthanized rather than waste valuable passenger space for pets during an evacuation. However, this evacuation was not wartime, and everyone hoped for a quick return to the base, so putting the animals down was never seriously considered. Taking care of the pets within the context of the evacuation, however, made life difficult for everyone. Many did not have leashes for their pet, and very few had pet carriers for the animals. The billeting rules forbade animals in the rooms. It was too hot to leave the animals in cars. There was no place to walk them and no place to put them. Cardboard boxes were brought from the commissary and base exchange, but there were so many pets the boxes had little effect. An emergency order went out to nearby bases for 2,000 pet carriers.[42]

Col Bruce Freeman, vice commander of Thirteenth Air Force, arrived later in the day. He found a mass of people jammed into the makeshift reception center. In an interview a year later, Freeman said, "About two o'clock in the morning, this young seaman, maybe 21 years old . . . wife looked to be about 19 or 20 . . . they were obviously just getting started in their married life. They came in and very quietly came up to me and said, "Sir, we don't have much, but whatever we have we will be happy to share. There are five rooms in our house, not counting the bathroom, we would be happy to take a family for each of the other four rooms."

"Well, I gotta tell you, I damn near cried – I cannot tell you what an inspirational experience it was . . . . It was Navy helping Air Force, and it long since lost service identity."[43]

A pregnant woman, who was at the head of the line and filling out paperwork, turned to her companion and said, "Now." Everyone knew what she meant, and someone hustled her out of the room into a hallway. Standing in the next line beside the woman was an obstetrician-evacuee. He followed the woman out and escorted her to Subic Bay hospital where her baby was born a little later.

The Mission Essential Team

At nearly midnight on June 10, the CAT commander, who had just returned from flying Clark's last F-4 to the bone yard, returned to his quarters in a taxi he had taken from Manila. He had the strange sensation of being the protagonist in a Rod Serling "Twilight Zone" episode. There were lights everywhere: street lights, porch lights, and lamps in the homes were all ablaze. No people. The streets and sidewalks were deserted. No cars. All the lights were on, but there was nobody home. What should have been a bustling community, even late at night, looked like a full-scale mock-up for a toy train advertisement.[44] There were no sounds. The air was dead quiet.

The next day, Colonel Anderegg, the CAT commander; Colonel Murphy; Colonel Grime; Colonel Dassler; and the few remaining other commanders refined plans to withdraw if the mountain erupted. The USGS had determined that the likelihood of a large eruption was a near certainty. The only question was how large.

They discussed three scenarios. A scenario in which an eruption would occur and send pyroclastic flows only to the northwest was judged to have a "high" probability; the occurrence of an even larger eruption, in which pyroclastic flows would cross a topographic divide and also flow to the east, towards the base, was judged to have a "moderate" probability. The possibility smaller eruptions might continue for weeks to years was also given a "moderate" probability.[45]

The stress on everyone was nearly palpable. General Studer had ordered the evacuation in the face of strong skepticism from Washington. USGS and PHIVOLCS were concerned that tens of thousands of Filipinos had not evacuated despite the warnings. The mayor of nearby Angeles City said, "they (the Americans) are overreacting" and "causing panic," comments that were published in both *The Washington Post* and *The New York Times.*[46] Everyone, commanders and scientists alike, was worried about the security policemen who were deployed across the base guarding homes and government property.

Colonel Dassler, and his deputy commander, Col Art Corwin, had decided to fight the way they had trained. In other words, they elected to use a system of defense that security forces had designed to protect the base from enemy attack. The system was called Air Base Ground Defense

Cars lined up on the aircraft parking ramp to receive last-minute evacuation instructions en route to Subic Bay.

(ABGD), and it was one they practiced countless times in preparation for an attack on the base from enemy forces.

The ABGD plan divided the base into defensive sectors. Each sector had a commander at a small command center that could communicate orders to his troops. Each sector reported to a central battle staff, which in turn received information and taskings from the main command post.

A young security police lieutenant, Chris Bargery, was commander of Alpha Sector. He set up his headquarters in the abandoned Wagner High School with 175 police under his command. Trucks, bricks, weapons, and other equipment were assigned to him so that he could operate independently and efficiently. Bargery, whose wife and two young children were at Subic Bay, moved out of his house and into the principal's office. He slept in a large, luxurious leather chair behind the principal's desk.

With only minor modifications, the ABGD plan fit perfectly into the plan to escape Pinatubo's wrath – if it came. The CAT commander would stay in the drum room with the volcanologists. If an eruption seemed imminent, the CAT commander would order "Hawkeye," the command post personnel, to blow the base siren for five minutes. When the ABGD commander heard the siren, he would direct each sector commander to withdraw his troops. This warning system had the benefit of being redundant since the troops guarding the base could get the warning to withdraw from either the bricks they carried or the siren. If either the brick system or the siren system failed, the other would provide the warning.

Once the alarm sounded, the mission essential team of cops, cooks, commanders and others would drive immediately to a tent city constructed at the far side of the base away from the volcano. If the eruption were very large, then the mission essential force would continue another nine miles to the Pampanga Agricultural College where Colonel Murphy and Colonel White's people had stockpiled supplies.

Subic Bay personnel bathe in the base swimming pool.

Implicit in the plan was enough warning of an eruption to move everyone to safety. Basic math revealed that a pyroclastic flow could reach the base in less than five minutes. No one wanted to contemplate the horror of a wall of superheated gases and ash slamming into Hill housing at 50 miles per hour.

The Alternate Command Post

Earlier in the month, Col Al Garcia, the base's communications commander, Colonels Murphy, White, and others had surveyed a building near Clark's east wall. The building had been a communications center during the Vietnam War, but was unused for the last several years. The building was an ideal location for an alternate command post.

During wartime, the siting of an alternate command post would be an elementary part of combat. If the enemy knocked out the primary command post, the commander or his deputy would still be able to direct the battle from the alternate.

The wily colonels had picked an ideal location. The building was as far away from the volcano as possible, yet still on base property. It had both a primary and a back-up electrical generator that powered the building. Both generators were equipped with air filters, and the air conditioning equipment was filtered, too. The brick system's central control functions were located in the building as well, and many of the base's telephone lines came through the building.

It was a large building, nearly empty of any furniture, and it had bathrooms and a few showers. The large, empty rooms could be used for many purposes: sleeping quarters, office space, supply storage or meeting rooms. The building sat just inside the gate that led off the base into the town of Dau (pronounced da-oo).

On June 10, the USGS-PHIVOLCS scientists moved their monitoring system from the Maryland Avenue site, which was on the southwest side of the parade field, to the Dau alternate command post. They did not want to move their equipment in the middle of an eruption, an act that would cause them to lose data while the mountain was erupting. The Dau site put them an additional five miles from the volcano. Although this move was not intended to overcome any lingering skepticism that an eruption would occur, it may have had that effect anyway.[47]

The beards set up the drum room on the west side of the building where windows permitted a clear view of the Zambales and Mount Pinatubo. They established a very crowded dormitory in two rooms behind the drum room. It was jammed with cots and the ubiquitous backpack each scientist carried.

Chaos at Subic

Down at Subic Bay, the previous day's chaos of finding shelter for 15,000 people had given way to confusion and disorganization. Colonel Grime had hoped that when he sent his people to Subic they would be billeted somewhat by organization. It did not happen. There was no organization of any kind. People were in the gymnasiums, schools, classrooms, hallways, chapels, homes, and any place they could find a roof over their heads to be out of the sun and recover from the exhaustion of the evacuation. Navy workers did their best to make the evacuees comfortable. Bubble wrap and cardboard boxes were provided as sleeping surfaces.[48]

While the colonels who had evacuated tried to figure out a way to establish a functioning organization, the first sergeants sprang into action. The "first shirts," as they are nicknamed, went to where the troops were and started accounting for each of them. They went into the dormitories, schools, and chapels. They drove the streets of the housing areas at San Miguel and Grande Island. As they worked, a list slowly grew of where people were camped. Colonel Goodman, who was acting as support group commander at Subic, was amazed at how thorough the ingenious first shirts were at finding their people. Slowly, agonizingly slowly, organization emerged and information started to flow to the evacuees.[49] Not all evacuees wanted to be found. Many approached the trip to Subic Bay like a holiday. Even the Navy hosts tried to entertain their guests by sponsoring a golf tournament, and organizing cook-outs with free, cold beer. Those who wanted to party did so, and military organization took a back seat.[50]

Colonel Freeman, the senior Air Force officer, acted to reduce the crush of humanity. Nearly 50 percent of the married enlisted members were married to Asian spouses, and about 80 percent of those were married to Filipinas. Freeman announced that anyone desiring leave who had a rotation date after July 1 could join their local relatives as long as the location was not within 30 kilometers (18 miles) of Mount Pinatubo. They were to return to Subic Bay no later than June 28.[51]

Navy officials assisted Freeman's efforts to reduce long lines at dining facilities by extending open hours to 24 hours a day at the base dining halls and by extending the open hours at other shopping and eating locations.[52] However, the Navy bases were swamped. The base exchange ran out of disposable diapers and dog food the first day.

The Vigil at Clark

The volcano, 25 miles away from Subic Bay, got uglier by the hour. A dome had started to extrude from the summit's northeast side. One of the tiltmeters, or inclinometers, started to tilt outward at an alarming rate. The drums were showing constant swarms of earthquakes, and the earthquakes were even closer to the surface. All evidence suggested that a shallow conduit was developing for delivery of magma to the surface. During a morning helicopter observation flight, observers identified a small (50-150 yards in diameter) lava dome on the northwest side of the mountain. The small dome was accompanied by the emission of a low, roiling ash cloud from the vicinity of the dome.[53]

Visual observation of the volcano was critical to the USGS-PHIVOLCS scientists. They had set up the drum room at PVO on Maryland Avenue so they could watch the volcano out the window while watching the drums at the same time. When they moved to the alternate command post, they again set up where windows faced the mountain. The helicopter flights in the base's UH-1s permitted close observation. However, there seemed to be no way to watch the volcano at night until someone suggested they use Cyclops.

Rube Goldberg would have loved Cyclops. The F-4Es, which had just departed the base, frequently carried a large, 2,000-pound pod on the belly of the aircraft. The pod contained an infrared sensor that could see targets at night. One of these pods had been mounted on top of the airfield control tower to assist security police forces in their continuing, and sometimes frustrating, efforts to interdict thieves who tried to sneak onto the vast reservation at night. A security policeman operated the pod from a small office in the control tower, scanning the perimeter wall and the large uninhabited areas of the base. If he spotted an intruder with Cyclops' infrared evil eye, he would vector security forces, frequently the base's mounted horse patrol, toward the intruder. Often, the intruders escaped, but Cyclops was a definite plus in the security forces' bag of tricks.

Cyclops had a clear line of sight to the volcano both night and day as long as no clouds were in the way. Soon, the volcanologists asked that a hot line be established between PVO and Cyclops in order to watch the mountain at night. Cyclops had one other interesting feature. It was rigged to a video recorder, also commandeered from the old F-4s, that could tape everything Cyclops saw.

The mounted horse patrol (MHP) flight of the security police group was the only unit of its kind in the Air Force. The MHP stables supported 20 horses in an area shared by recreation services' horses. Each evening, MHP riders would report to the stables, saddle up and patrol the base. Most of their patrol routes went through the elephant grass, creek beds, and banana groves of the most remote areas of the base. The riders always patrolled in pairs, carried weapons for self-defense and maintained radio contact with central security control. The horse frequently sensed intruders before the riders. When a horse stopped, pricked up its ears and looked in a certain direction, a smart rider gave the horse his head. Sometimes the riders wore night-vision goggles, but the extra weight and danger of losing the valuable goggles in the jungle hardly made the effort worthwhile since the horses saw nearly as well.[54]

The night of June 11, everyone at Clark who was not on duty slept in his own bed. Despite the mountain's volatility, General Studer believed a good night's rest was important for everyone. While Cyclops scanned the river valleys and the night watch of volcanologists leaned over the drums with their hands in their pockets, the mission essential force rested. It was the last peaceful night they would have for a long time. The next day would see them become the Ash Warriors.

NOTES TO CHAPTER 3

1 Fire and Mud, p 73.

2 Ibid., p 74.

3 Interview, General (Ret) Jimmie V. Adams with the author, January 20, 1999, audio tape at HQ USAF/HO, Bolling AFB, DC, [hereafter cited as Adams interview].

4 Author's journal.

5 Ibid.

6 Fire and Mud, p 77.

7 Murphy interview.

8 Murphy interview, sup doc Rand interview.

9 Murphy interview; Rand interview; Newhall e-mail.

10 Rand interview.

11 Philippine Flyer, "Just the Facts . . . . ," June 7, 1991, p 5.

12 Rand interview.

13 Ibid.

14 Murphy interview.

15 Rand interview.

16 Newhall letter, 27 July.

17 Ibid.

18 Grime interview 1992.

19 Adams interview.

20 Fire and Mud, p 80.

21 Anderegg, pilot log book.

22 Murphy interview.

23 Hoblitt notes, p 17.

24 Fire and Mud, p 80.

25 Philippine Flyer, June 7, 1991, p 5.

26 Fire and Mud, p 80.

27 Ibid.; Rand interview.

28 Studer interview.

29 Hoblitt notes.

30 Fire and Mud, p 80.

31 Interview, David H. Krieger with author, February 4, 1999, audio tape at USAF/HO, Bolling AFB, DC [hereafter cited as Krieger interview].

32 Interview, Colonel Bruce Freeman with PACAF/HO, October 9, 1992 [hereafter cited as Freeman interview].

33 Murphy interview; Rand interview.

34 Calendar.

35 Rand interview.

36 Ibid.

37 Grime interview 1999.

38 Goodman interview.

39 Krieger interview.

40 Ibid.

41 Interview, Barbara Collins with author, October 30, 1998.

42 Goodman interview.

43 Freeman interview.

44 Author's journal.

45 Fire and Mud, p 81.

46 Ibid., p 80.

47 Ibid., p 81.

48 Krieger interview.

49 Goodman interview.

50 Krieger interview, Goodman interview.

51 Philippine Flyer, special edition, June 11, 1998.

52 Ibid.

53 Fire and Mud, p 9.

54 Interview, MSgt Harlan E. Mikkelson with author, October 23, 1998 [hereafter cited as Mikkelson interview].

Chapter 4 FROGS IN A COOKING POT

The volcano erupted at 8:50 a.m., June 12, 1991. There was no warning except a frantic call from the command post, Hawkeye, on the brick, "It just blew! The volcano just erupted!"

The crisis action team was gathering in the main command post for a 9:00 a.m. meeting. When he heard the call on the brick, the CAT commander used his brick to initiate the withdrawal of mission essential personnel to the far eastern side of the flightline, "Hawkeye, this is Lucky, blow the siren."

Dave Harlow, the USGS scientist who had replaced Chris Newhall,* was at the command post for the upcoming meeting. He, Lucky, and the rest of the CAT sprinted down the hallway and out the front door of the command center. Before they got halfway down the hallway, they knew the withdrawal plan was underway because all the lights went out as the power plant workers "pulled the plug" on the huge generators to protect them from possible ash fall. Almost instantly, the hallway emergency lighting kicked on, and the sprinting group was saved the embarrassment of running into each other like some Keystone Cops movie. The CAT commander made a mental note to tell Mr. Jim Weed, a power plant supervisor, to delay cutting the power so people could get out of the buildings without broken bones.

The group emerged into a bright, clear perfect day and stopped to look westward toward the Zambales. They, as every other person there that day, stared in amazement. A huge plume of gray, roiling ash was rocketing upward out of Pinatubo into the cloudless, blue sky.

In the same instant it was terrifying and awesome. The column continued to climb at an astonishing rate. It looked very much like the movies of nuclear test explosions – a huge, mushroom-shaped cloud. The base's weather radar could see the ash cloud and tracked its top. Ten miles high. Eleven miles high. Twelve miles high![1] Later measurements would show that the plume climbed at 400 meters per second, or 72,000 feet per minute.[2]

The top of the mushroom started to spread out in an enormous umbrella as the upper level winds started to diffuse the ash. There was no sound above the steady wail of the warning sirens.

Colonel Rand was in General Studer's office while the Thirteenth Air Force commander was speaking on the telephone with another flag officer. Rand recalled, "Power went off, sirens started going, General Studer's brick beeped, and he said, 'That's it, Mark, we're out of here,' and jumped up and got his hat and ran outside, looked up over my building and everybody who was there in that area was racing for a car. There wasn't anybody taking their time. As I looked up I could see that plume come up above, I knew where the mountain was, we all knew where the mountain was. It was right behind my building. I didn't see anything, then all of a sudden . . . this plume . . . was going up so fast I couldn't believe it. I watched it and . . . jumped in the car with the protocol officer, Spuds McKenzie, and we got down on the flightline . . . to the far side of the base. Cars were going everywhere, they didn't know [where] they were supposed to be going. I'd seen the evacuation route, so we stopped . . . I told Spuds to go get that line of cars that's down there and head them back this way. I just started waving cars out towards the right area. All manner of vehicles, every kind of vehicle you've ever

* *Newhall had returned to USGS headquarters in order to coordinate processing of data from the volcano.*

The wait and the uncertainty was over as Mount Pinatubo finally erupted hurling tons of ash and debris over 12 miles into the air.

seen and people in every state of disarray and undress you've ever seen. I mean it looked like if they were in the shower, they'd put a towel around them, jumped in the car, and headed across the flightline."[3]

Harlow and Lucky discussed the threat. Harlow saw no pyroclastic flows (PFs) and was not very concerned that this eruption presented much of a threat to the base. He called it a photo-op. Lucky, who was concerned about the need to send the mission essential force further away to the Pampanga Agricultural College, asked if this was "the big one."

"No," Harlow replied, "this is probably just the volcano clearing its throat." He further explained that they would now have to take a close look at the data to determine how much, if any, pressure this eruption had released.

...this is probably just the volcano clearing its throat

As they spoke, myriad vehicles were whizzing past them and out onto the flightline. One pickup truck zipped past them, its driver steering with his left hand as he used his right to aim a video camera out the window and behind him to tape the spectacular eruption. Many airmen took video pictures of the eruption as they ran from dormitories and duty locations to POVs or government vehicles to respond to the siren's wail. The tapes are filled with expletives, exclamations and nervous laughter as doors slam, engines race and tires squeal in their haste to get across the runways to the relative safety of the far side of the base. Amazingly, no accidents happened despite the helter-skelter of what seemed to be total chaos. The CAT had not directed a practice of the evacuation procedure.

Those who drove down Wirt-Davis Avenue passed a large sign in front of the officers' club. It announced the abandoned club's plans for a social event scheduled for the coming weekend. In bright red letters, the club proclaimed a "Last Days of Pompeii" party was scheduled for the upcoming Friday, June 14. One could assume the proper attire for the party to be a toga.

Gallows humor... A sign in front of the Clark Officers' Club announced a party that was planned before the situation became serious.

In accordance with the "bugout plan," the security police group assembled near the Dau command post, but most of the remainder of the mission essential force, about 600 people, continued to the Pampanga Agricultural College. The purpose was to get as many people as far away from the volcano as prudent while keeping the security forces nearby so that they could return to security duties as soon as possible.[4]

One of the young security policemen at Clark was SSgt John M. Thomas of the 3d Law Enforcement Squadron. He recalls, "We were playing basketball by the dormitory when the siren went off. I'm not sure if it went off before or after we saw the cloud, but they happened at about the same time. It looked like an atomic explosion, a mushroom cloud that was going up into the sky."[5] Sergeant Thomas did not own a car, so he went to the nearby armory that had been designated as a rally point. He and others who did not own cars loaded onto an evacuation bus that took them to the Dau command post across the runways.

Harlow and Lucky were soon joined by General Studer, Colonel Grime, and Colonel Rand. Harlow filled them in quickly about what he thought was going on with the volcano. Both Studer and Grime were greatly relieved that the volcano had erupted but was not causing any damage or threatening the base.

When General Studer ordered the evacuation of dependents and non-mission essential service members, only two days earlier, he took several big risks to his credibility as a commander and a leader. In his view, losing his subordinates' trust was the largest risk. If he ordered an evacuation, yet nothing happened, the men and women of the base would be very reluctant to cooperate with any plan to re-occupy and possibly re-evacuate the base. Neither Studer nor Grime wanted to establish a cycle that whip-lashed their people back and forth. Of course, the skeptics at the headquarters grew more skeptical as a function of their distance from the volcano. So, as they watched the eruption spread out in the upper atmosphere, both felt a sense of relief that their judgments had been sound. The evacuation process had been difficult for everyone, but it had been the right thing to do. The proof was there for all to see, an enormous tower of ugly, gray ash hanging over the base.[6]

Regrouping

Gladiator called Lucky on the brick to ask how long it would be before the security forces could resume their Air Base Ground Defense posture. Some of his units had reported intruders coming over the wall as soon as the evacuation started. Apparently the thieves knew what the siren meant, too. The CAT commander told him to resume immediately since the eruption did not seem to be a threat to the base. The security forces' training had prepared them for retreat, regroup and counter-attack operations, so the security police group, already regrouped and ready to move in, bolted back into the housing area. Looters scattered in every direction as the security forces caught them completely off guard. The thieves did not have enough time to do much damage. One, who had bitten off more than he could chew, was caught hauling 10 bicycles in a pull cart.[7]

The CAT commander and Eagle, the wing commander, drove over to the alternate command post near Dau and discussed the situation with the USGS-PHIVOLCS team. They had not passed a warning because they had only a few seconds indication from the drums and the RSAM that an eruption was occurring. Although energy levels in the mountain had decreased somewhat, it was clear to all the scientists that the eruption was not the big one. Indeed, it was just as Harlow had guessed – a photo op of the mountain clearing its throat a little. The beards were somewhat embarrassed that they had not seen the eruption coming. Eagle and Gator were very glad they had taken the conservative approach and cleared the base. Nobody wanted to contemplate the possible chaos the spectacular eruption would have caused if the base was crowded with its normal population.

A Bad Decision

Perhaps the biggest scare of the morning's eruption came because of a well-intentioned decision that nearly went wrong. Many of the evacuees at Subic Bay were coming to the end of their tours in the Philippines and had port calls to depart Manila during the month of June. In order to relieve the congestion at Subic, Air Force officials at Subic and at personnel headquarters in the United States had agreed to accelerate the departure of these "short timers," by moving up their departures to an as-soon-as-possible status. In order to give these early departees an opportunity to retrieve personal items from their homes at Clark, a scheme was devised. The people would come to Clark on two buses that would drop them off at their homes. They had one hour to pack anything they wanted, then the buses would pick them up and return them to Subic. The idea was to allow a few people at a time to get back into their homes to retrieve items they had forgotten, or did not take because they had not planned to leave for the United States having never returned to their homes. Goodman and Freeman, the senior officers at Subic, were already discovering that many families had taken few or none of the items on the evacuation checklists. At Clark, security policemen and transportation specialists devised a plan, under the supervision of the CAT commander, to get the people into and out of their quarters quickly, with no danger of anyone being left behind if the volcano blew while they were in the houses packing suitcases. As people were dropped off, an orange traffic cone was left in front of the house so that the bus could pick them up quickly. Sure enough, just as the last person was dropped off, the volcano erupted. The bus driver quickly retraced his steps and loaded frightened evacuees back on the bus as the siren wailed and the huge ash cloud grew ominously overhead. The plan to get folks back into their homes was abandoned.

Improving the Bugout Plan

Later that morning, the CAT updated plans to improve the flow of the mission essential team out of the housing areas, down to the flightline and across the runways to the tent city beside the Dau alternate command post. The morning "bugout," as they came to be called, had been successful but had many rough edges. The most obvious rough edge was that not everyone understood the plan. Two security policemen not only did not stop at Dau; they did not stop until they got to Manila. When they called their supervisor for instructions a few hours after the eruption, he told them in no uncertain terms how they should come back to Clark and exactly how much time they had to do it.

Many drivers that raced out onto the flightline did not know how to get to the other side of the base. Some were so ingrained with the pre-volcano prohibition against driving on the flightline that they milled up and down the street rather than go into what had been a restricted area.

Another problem was that there was no lighting on the flightline to show drivers the way at night. Aircraft maintenance troops spent the day bringing "light-alls," trailered floodlights used for nighttime aircraft maintenance, to strategic spots along the evacuation route.

By late that afternoon, CAT members had re-briefed the plan to their troops so that everyone would know where to go if the siren blew again, and how to get there. No one would be allowed to sleep at homes in the areas nearest the volcano. Approved sleeping areas were the dormitories near the flightline, the tent city beside the Dau command post, and the Dau command post itself. A buddy system was initiated so that no one traveled, ate, slept, or worked on the base alone. The CAT did not want to have anybody asleep where they might be left behind in a bugout. Anyone who was on the volcano side of the flightline had to have an operable brick that was always on and immediate access to a vehicle. In fact, no one was to cross to the volcano side of the flightline without his/her commander's permission and specific reason to be there.

Col Al Garcia, the 1961st Communications Group commander, and his people had been working feverishly for several days to upgrade the communications link in the old building, and it was ready for full-time use. So, Lucky decided to move the entire command post operation (Hawkeye) over to the alternate location at Dau so that all command and control functions, as well as the drum room, would be co-located. The move would allow the CAT commander to be with the beards constantly, yet have all the command and control functions provided by Hawkeye a few steps away.

All power lines and insulators had to be washed clean of ash everytime ash blew across the base for weeks after the main eruptions.

By 4:00 p.m., the transition to the Dau command post was complete and everything there was operational. Hawkeye was in place and the brick network was complete. The only glitch was that the siren control had not been moved yet, so someone from Hawkeye had to stay at the command post to push the button if the CAT commander ordered it done. Maj Hugh Riley, who had flown the last F-4 back to the bone yard with Colonel Anderegg, the vice-wing commander, had returned and resumed his duties as Hawkeye commander.

The volcanologists made it clear that the morning eruption was a precursor of others, perhaps bigger.[8] Harlow, Ewert and the others were in their "worry mode." They paced back and forth in front of the drums and occasionally stopped in front of the drums, stuffed their hands into their pockets, bent at the waist, and peered at the squiggles. They were concerned about not getting much warning about the eruption but not surprised. In just a few seconds, the earthquakes had increased from short, small traces on the drums to wild, constant, side-to-side swings of the needles on every drum. The eruption had continued for approximately 38 minutes.[9] Although PFs did not appear on the Clark side of the volcano, some observers saw small ones on the northwest side of the volcano.

None of the ash fell on the base. However, winds from the northeast, unusual for the time

of year, carried it toward Subic Bay and its auxiliary housing area, San Miguel, a few miles to the north. Surprised residents and evacuees watched it fall like a light snow until everything was covered with a very thin dust. Some quickly discovered it could not be brushed away easily; the stuff was abrasive and instantly scratched hoods, roofs and trunk lids.

All afternoon and evening of June 12, preparations continued to stockpile food and water at the Dau command post and further west at the college. One of the bigger concerns was that ash fall would short out the base electrical system thereby rendering useless the water pumps that replenished the water storage tanks. After an afternoon CAT meeting, an Air Force fireman approached the CAT commander and said the fire department had some large tank trucks that all together held many thousand gallons of water. Further, the fireman suggested, if the tanks were rinsed, then filled with fresh water, he thought the water could be chlorinated and stored right in the trucks. The CAT commander asked him to talk to Lt Col (Doctor) Green, and if it would work to go ahead with the idea. Most everyone who heard the brief side discussion immediately forgot about it – except the fireman and Doc Green, the medical officer who was part of the mission essential team along with a few of his corpsmen.

Getting Out the Word

By nightfall, the telephones had stopped ringing off the hooks. Inquiries from around the world poured in all day. Of course, the headquarters at both PACAF and the Pentagon wanted to know the situation and what actions were planned. There was mild interest from news media in the United States, and all such calls were referred to Mercury. Rumors were already running rampant throughout the evacuee population at Subic Bay that many homes were looted; some even thought the base had been abandoned completely. Colonel Rand made arrangements with his troops at Subic to start publishing an edition of the *Philippine Flyer* at Subic to keep everyone informed about the true happenings both at Clark and Subic Bay.

Essentially the mission essential force was cut off from communicating with their families in the United States. The telephones were operational, but since no one could go home to call, commercial phones were of no use to them. Military phone lines were clogged with official business calls. Wizard's communications group started advertising the availability of "MARSgrams"* around the base. Anyone who wanted to get a message home could fill out a form that looked similar to a telegram blank. Garcia's 1961st communications people then collected the forms and took them to the MARS station where they were read over long-range radio to a station near the sender's home. The receivers on the stateside end would then make a written copy of the message and deliver it to the airman's home address.[10]

Mail was another problem for the mission essential force since the postal squadron was operating with only a skeleton crew. Outgoing mail was limited to letters only, and incoming mail was a nightmare to sort since most people were at Subic and it took hours to extract mail for those who were still at Clark on the mission essential force.[11]

Earlier in the afternoon, the 600 support personnel who had evacuated to the college returned to their duties at the base. The CAT commander and wing commander agreed that the next time a bugout was needed, all mission essential personnel would wait in the Dau area and near the gates for a second directive to move everyone out to the college. The bugout in the morning had gone fairly smoothly, so it seemed reasonable rather than splitting the force. With the experience of one eruption under their belts, the mission essential team was more confident of its ability to move quickly if a more severe eruption happened.

A Second Eruption

By 10:00 p.m. those not on duty in the Dau post were asleep. The CAT commander had turned over his drum watch to Lt Col Al Shirley, call sign Thruster. Shortly thereafter the volcanologists on duty noticed several long periods of earthquakes. They told Shirley that an eruption might be imminent, so Thruster awoke Lucky from the "VIP Suite," a small room at the back of the command post with three army cots that had been reserved for General Studer, Colonel Grime, and the CAT commander.

Thruster, his voice urgent, related that the volcano looked like it might erupt again, but they could not see it in the dark. The CAT commander picked his brick up from the floor and transmitted, "Hawkeye, Lucky, blow the siren!" The airman on duty at the abandoned main command post immediately turned the siren on, and it started its ear-splitting wail across the base. By the time Lucky could walk across the building to the drum room, cars were already streaming out of the dormitories across the field and across the runways. This evacuation seemed more orderly than the morning event as the cars and trucks all moved smoothly through the cordon of light fixtures that showed the way.

All the scientists were clustered in front of the drums, several in the hands-in-pockets-bent-over mode. Their main concern was that they could not see what was going on. Thruster picked up a nearby telephone and dialed the airman who was operating Cyclops, the jury-rigged infrared fighter pod that was mounted on the airfield control tower. The airman already had pointed Cyclops in the direction of Pinatubo, so he reported what he saw. The eruption looked just like the earlier one, and he saw nothing coming down the mountain range toward the base. Everyone in the drum room breathed a bit more easily.

The weathermen, who were monitoring the base weather radar, called in and reported that the eruption was going even higher than the morning one. Ultimately, it topped-out nearly 15 miles high.[12]

Once again, the volcanologists explained that this eruption was not the climactic event. The RSAM still indicated an extremely high energy level under the mountain. These throat clearings, although spectacular, were relatively inconsequential, in their opinion. In general, the mountain had returned to its pre-eruptive levels of earthquakes and energy level. The CAT commander returned the security police to its duties, and those not on duty went back to bed. However, the CAT passed another directive through the chain of command to all personnel. Wherever anyone was asleep there must also be someone awake. The CAT was concerned that ash fall might disable the siren warning system. If that happened, the bricks would be the only way to pass the word to bugout, and a single system was clearly inadequate considering the stakes.

* *Military Affiliate Radio System, a military-sponsored ham radio net.*

Although the CAT directives at this time only required those who lived in the western housing areas to sleep near the flightline or the Dau command post, the security group commander noted that, not surprisingly, more beds were filled in tent city following the second bugout. The military men and women of Clark Air Base were experienced troops, and experienced troops understand that one sleeps where and when one can. Clearly, many had decided that if it took sleeping on a cot in tent city to get a good night's sleep then so be it.

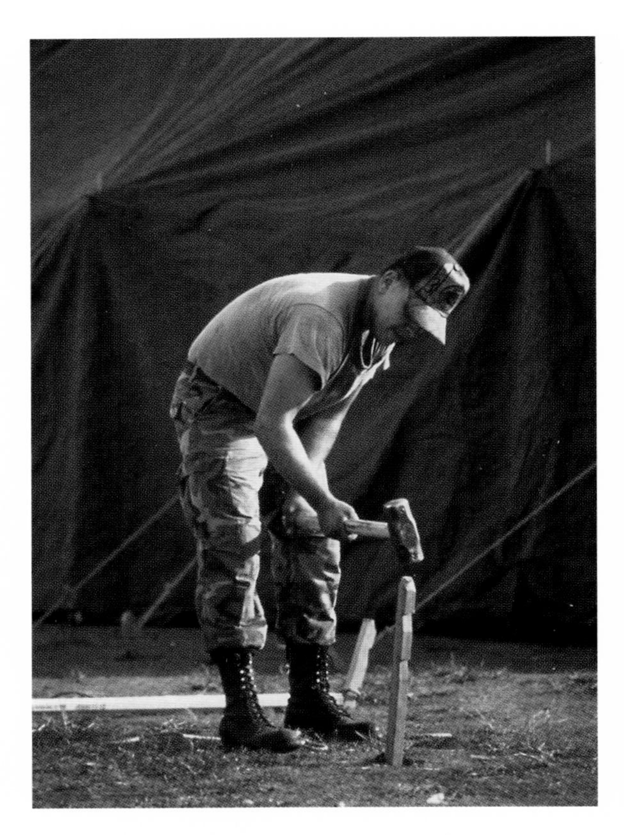

Personnel erected a tent city in order to be able to sleep nearer to their evacuation point.

June 13, 1991 dawned as another beautiful dry-season morning. Gray ash covered the crest and flanks of Mount Pinatubo, and a continuous light gray plume emitted from its summit. At a meeting of unit commanders that morning, Eagle, 3 TFW commander, congratulated everyone on "doing a great job of keeping people safe and protecting our property. We have a very important job here, and I'm proud of the way we're all pitching in to do it right."

He further emphasized that, "we'll probably have to respond to more eruptions for the foreseeable future, and . . . there really isn't any way to tell when – volcanoes are just too unpredictable. For that matter," he continued, "we can't even say how long this situation may go on, so we need to make certain that everyone gets some quality rest, so they can continue to take the alarms seriously and respond according to the plan until the volcano subsides. Things are going very well, and I like the way we're pulling together as a team. I'll tell you, that old Clark pride is really evident."[13]

Another One

By 8:30 a.m. the same earthquake patterns were appearing on the drums that had been seen before the previous evening's eruption. At 8:41 a.m., while the volcanologists and the CAT were discussing the possibility of another eruption, the volcano erupted for its third time. Hawkeye sounded the siren and another bugout began. This time, though, the flow of traffic across the flightline was orderly and more efficient, and all personnel were out of the western end of the base in less than five minutes.

An eruption early in the sequence of blasts. The unusual Cassandra's Rings were only seen on this eruption and were probably the result of heat, temperature inversions and the speed at which the column ascended.

Again, it was a towering ash plume that rocketed up to nearly 15 miles.[14] This eruption looked even more like a nuclear explosion as the rapidly climbing ash plume hit the tropopause* at approximately 8 miles in altitude. Once the hot ash plume hit the rapidly cooling air of the tropopause, large concentric rings started to form around the head of the mushroom. The effect was at once stunning and horrific. As soon as the scientists were certain no PFs were observed or likely, Lucky returned the security forces to their posts and others to their support duties.

Many noticed that the three eruptions had occurred at nearly 12 hour intervals. Lucky discussed this with the beards. He wondered if the volcano were showing signs of becoming an "Old Faithful." The scientists were unimpressed with his theory and dismissed the timing as a coincidence.

Some Philippine newspapers reported that weapons, especially nuclear weapons, stored at

* The tropopause is the boundary between the upper troposphere and the lower stratosphere; it varies in altitude from approximately 5 miles at the poles to approximately 11 miles at the equator.

Clark were in danger of accidental detonation from lava flows. *The Guardian*, a Manila paper carried the headline, "US issues nuclear alert as volcano gathers strength," and alleged that the US issued a "nuclear alert . . . . After servicemen were forced to abandon Clark Air Base in the Philippines following another series of huge explosions from the Mount Pinatubo volcano." The article also quoted a US official saying, "we are abandoning the base to nature." The article was reprinted in a number of Manila papers.

The US Embassy was quick to rebut the article and issued a statement, "The US Embassy categorically denies allegations which first appeared in *The Guardian* and later disseminated by the local press that the US has issued a 'nuclear alert' at Clark Air Base. No such alert was ever contemplated or issued. The Embassy wishes to emphasize that no US weapons system at Clark or elsewhere in the Philippines is endangered by the ongoing eruptions of Mount Pinatubo."[15]

...the "water" was getting warmer by the day

Indeed, the only weapons stored at Clark were some 3,370 short tons of conventional munitions stored in two munitions storage areas (MSAs).[16] The MSA nearest Pinatubo was empty. The MSA nearest the Dau command post contained some munitions that were considered "sensitive" because they would be useful to terrorists if they could get their hands on them. These sensitive munitions were Stinger* missiles and plastic explosives.

By the early evening hours of June 13, the volcano was again showing the signs of an impending eruption, and the volcanologists issued a warning to the base and to the surrounding area.[17] The CAT commander blew the siren and the fourth bugout in 36 hours was underway. Each unit reported in that they were standing by the gates and ready to move out to the college. The beards watched the drums. The CAT watched the beards. Nothing happened. The earthquake activity gradually subsided until it was at its previous level. By the time it was dark, the CAT had returned everyone to normal duties.

The efforts of the security police group to return quickly to its patrol duties were effective despite the increase in intrusions by thieves during the bugouts. Once the housing areas, work areas, and main storage areas were devoid of dependents and non-essential personnel, the cops' job of identifying intruders got easier. If they saw someone in the housing area, for example, who was not wearing BDUs,† then that someone did not belong there. Also, there were many more patrolmen on duty since shifts were longer and more frequent, and the troops did not have other ancillary duties, such as training meetings, to attend.[18] During the previous three eruptions there were only two confirmed burglaries within the housing areas.[19]

The volcanologists and Lucky, the CAT commander, sat in the drum room and shared packages from MREs.‡ John Ewert, who had shown an eerie ability to guess what the volcano would do, related a metaphor. He said, "You know, if you drop a frog into boiling water, he'll hop right out to save his skin. But if you put a frog into cold water, then gradually heat it to boiling, the frog will sit there until he's cooked." The assembled group of scientists, officers, and sergeants laughed, but everyone got the point. The water was getting warmer by the day.

By the morning of June 14, the Dau command post looked like a fraternity house on Sunday morning. Sleeping bodies filled nearly every room. Some were on cots while others slept in mummy bags on the floors, tables, and any place there was enough room to stretch out. Gladiator reported that the tent city was nearly full. Many people slept in their cars, including Mercury.[20]

Several corners of the rooms were stacked high with MREs, cases of soda, bottled water and snacks. Paper coffee cups littered the drum room along with empty containers of the CAT/volcanologists' snack of choice – Cheese Puffs. CAT members joked about getting permanently orange fingertips from chowing down on the tasty treats.

Mother Nature Ups the Ante

At the morning CAT meeting, some bad news came from Maj Bill Nichols' weather forecasters. A typhoon, named Yunya, was headed for Luzon, Clark's island. A typhoon is the Pacific version of a hurricane, and this one packed winds of over 100 miles per hour. The winds were not a big concern because a large mountain ridge along the east edge of Luzon usually protected Clark from intense typhoon winds. The real worry was the attendant rainfall, and Yunya carried very heavy rains.

** The Stinger is a small, heat-seeking missile carried in a short, lightweight launching tube. One person can fire it, and it is capable of bringing down even large aircraft.*

† Battle Dress Uniform, a newer version of fatigues.

‡ Meals Ready to Eat, the modern version of the infamous WW II C-rations. A single MRE box contains several vacuum-sealed plastic bags that contain a variety of food items. It is common for diners to open a box each, then trade individual packets until one gets the things he/she prefers.

As the weatherman spoke to the CAT, the room seemed to deflate slowly. The look on each face was "What next? A cloud of locusts?" Most were aware of the devastation wet ash could cause, and several recalled the example the scientists used about the glass of dry sand being filled with water.* The three earlier eruptions had put nearly an inch of ash on the ground at San Miguel, the Navy auxiliary housing area that held nearly two thousand evacuees. One inch was not a problem, but the mountain showed no sign of reversing its course of continuing eruptions.

The CAT set about making plans to deal with Typhoon Yunya. COR 3† was declared. The tent city could not withstand typhoon winds, so an alternate plan was devised. If the siren blew after 6:00 p.m., troops would go to two large hangars on the flightline to take shelter from the typhoon while waiting to see if further evacuation to the college would be needed. The CAT distributed a map to show how to find the hangars and gave instructions for the use of the hangars. A copy was given to every mission essential member, and it was duplicated in an edition of the base newspaper.[21]

Sweltering in the Sun

At about 11:00 a.m., the earthquake activity in the mountain was increasing in both frequency and amplitude at an increasing rate. Lucky, confident he could recognize an impending eruption on the drums and the RSAM, elected to direct a bugout. He radioed, "Hawkeye, Lucky, blow the siren."

As the siren sounded, Maj Hugh Riley's voice came on the net, "Attention on the net. Attention on the net. Mount Pinatubo is erupting, all personnel proceed immediately to the gates. I say again, Mount Pinatubo is erupting, all personnel proceed immediately to the gates!" Riley was on duty again. He and Capt John Cherry always seemed to be available no matter the time of day or night. All the command post controllers were dedicated officers and sergeants who were Johnny-on-the-spot whenever the CAT needed something done. Of course, the mountain was not erupting, but Lucky said nothing since Riley's radio call had the desired effect. Cars and trucks were streaming across the airfield to marshal at the gates.

"I say again, Mount Pinatubo is erupting, all personnel proceed immediately to the gates"

An hour went by and nothing happened. Ewert leaned over the drums and peered out the window. He turned to the CAT commander and grinned, "I don't think it's going to blow for another hour or so."

Dassler's gravely voice came through the brick, "Lucky, Gladiator. It's hot as hell out here in these vehicles. How 'bout getting some water out here to us while we wait."

"Wilco,‡ Gladiator," Lucky responded.

Within a second, an unfamiliar voice came on the net. "Lucky, this is Personnel Bobbitt, I'm on my way out with a truckload of sodas."

"Roger, thanks," Lucky answered.

The CAT commander wondered aloud, "Who the hell is Personnel Bobbitt?" Whoever he was, he was using the standard radio call sign the Air Force designates for those who do not have one. Standard procedure prohibits the use of rank on the radio, so those who do not have an assigned call sign precede the last name with the appellation "personnel." It sounded silly, but Bobbitt was doing what he had been trained to do. The services squadron representative on the CAT revealed that Personnel Bobbitt was one buck sergeant Jim Bobbitt. The services squadron was responsible for operating the dining halls and living quarters.

The convoys sat at the gate for another hour in the sweltering tropical sun. As canteens went dry and Bobbitt's truckload of sodas emptied, tempers grew short. The services squadron commander came on the net, "Lucky, Services 1, if you'll let me and a small squad go back across the flightline, we can prepare a hot lunch."

Lucky answered, "Go ahead Services 1, but make sure you have the vehicles pointed out and stay on the net."

Finally, after nearly two hours in the hot sun, the CAT commander released the convoys to return to their duty stations, "Hawkeye, Lucky. Bring everyone back to duty stations."

The CAT and the scientists watched as the stream of vehicles started back across the runway, bumper-to-bumper. Most of them were private vehicles. Many of the troops had their cars packed with treasures from home: wedding albums, kids' first toys, video tapes of first birthdays and the others from which lifetime memories are made. They were not going to leave Clark, even for a moment, without those treasures.

* *See Chapter 3.*

† *Typhoon Condition of Readiness, a warning system used by military forces in the Pacific. Level 3 means winds of 50 knots or higher expected within 48 hours; COR 2 means winds of greater than 50 knots in less than 24 hours.*

‡ *Standard radio terminology for "I will comply."*

Ten minutes later, the volcano erupted, and the mission essential force went through another bugout. The CAT held the convoys at the gate while Pinatubo threw ash and steam 12 miles into the sky. This eruption was not as spectacular as previous ones because high clouds, precursors of Typhoon Yunya, blocked the view.

And Another

Two hours later, another eruption came. CAT members wondered how long the volcano could maintain the pace. Surely, they wondered, it must be close to exhausting itself. The volcanologists pointed to the drums and the RSAM. By this time, the RSAM was five times higher than the level that had prompted General Studer to evacuate the base four days earlier. The energy level in the mountain inexorably continued to build, stepping ever higher with each eruption. During the night hours of June 14, the volcano erupted three more times.

"...one of the only times I ever feared for my life because of the unknown"

Volcanology 201

Every evening at 8:00 p.m., the USGS-PHIVOLCS team met to discuss Pinatubo. At the evening meeting of June 14, there were two particularly bad pieces of news. Typhoon Yunya was only 12 to 15 hours away from Clark. Even more disturbing was the news the team had just learned from Newhall back in the United States. Satellite photographs of Mount Pinatubo, and the analysis of those pictures and other data, suggested that the magma field stimulating Pinatubo's eruptions was estimated to be 3 to 5 square kilometers. The scientists offered a perspective: the magma field that powered the Mount Saint Helens blast was about one square kilometer. Further, they offered, the difference was a power of the square. Three times the magma involvement meant the possibility of an eruption nine times larger.

On the positive side of the scale, though, was the fact that Clark was on the "good" side of the volcano. The vents, explosions, and faults were all facing away from Clark to the northwest and west. The three-dimensional plots of the earthquake activity clearly showed the involvement to be on the side of the mountain away from Clark. Any large blast should go away from the base. Subic Bay and the evacuees were well outside any danger zone of even the largest possible explosion.

On the negative side of the scale, several Air Force officers noticed that the scientists had packed their knapsacks and were keeping them close at hand. Not a good sign at all.

All Nighter

One-by-one the CAT and scientists wandered out of the drum room and headed for a place to sleep. Lucky, Thruster, Huge (Major Riley), and geologist Rick Hoblitt sat in the drum room to discuss the situation. They pounded coffee, nibbled on Cheese Puffs and scavenged dinners from partial MREs that had been left from earlier diners.

Gladiator came to the drum room. He said, "We've had three bugouts in the last three hours. Everyone out there is exhausted. Let's let them sleep. The dorms are only two miles closer to the mountain than we are. Every time you blow the siren fewer and fewer actually respond. More and more, they're deciding to take their chances with the mountain just to get some sleep."[22]

Lucky agreed. He had no doubt that someone would get killed if the night bugouts continued. Someone would fall down the stairs or get run over by a speeding car in the dark. Both Eagle and Gator were sleeping in the "VIP Suite," and the CAT commander did not wake them up. Everyone was exhausted to the point where individual good judgment was seriously in question, and Lucky figured Eagle and Gator needed to be fresh by the time the sun came up. Ultimately, they were responsible for everything and everyone.

Sergeant Thomas recalls the fatigue, "The time period is rather vague, but the thing I remember the most, and it still sends chills down my spine, is the siren going off so much. There was no way you could sit back and say, 'Well, it's just the siren going off again,' because you just didn't know [what could happen]. I recollect many times when I thought I was getting some sleep. It was so tiring because when you were off [duty] you just didn't get any down time. It was exciting. It was exhilarating. But it was, at the same time, one of the only times I ever feared for my life because of the unknown."[23]

At about 2:00 a.m., the young airman monitoring Cyclops, called the CAT to report what appeared to be a small PF on Clark's side of the mountain. He had video taped the event and wanted to know if the volcanologists wanted to see it. Hoblitt immediately responded in the affirmative, and a runner was dispatched to retrieve the tape.

The tape showed what looked like a small, thin PF curling around the south side of the mountain about a thousand feet down from its summit. Hoblitt looked at the tape over and over. It was difficult to orient oneself to the tape because the Cyclops infrared pod had to rotate 180 degrees as it swung toward the mountain, so the picture was upside down. Also, the visibility was poor due to Typhoon Yunya's encroaching clouds.

Rick Hoblitt finally threw in the towel and went to bed after puzzling over the tape for nearly two hours. Only Bigfoot, one of the beards, Thruster and Lucky remained in the drum room. Suddenly, Hoblitt raced out of the beards' sleeping room and announced in an excited voice, "It's not a PF! It's a crack in the top of the mountain!"[24]

He hurriedly put the tape back in the machine and paused the tape at the best view. Using a topographical chart to help him, he made his point. The whole top of the mountain was loosening from the pressure below much like a cork in a champagne bottle.

"See," he said, "there's no gully or river there for a PF to flow through. That's what's wrong that I couldn't figure out. The only thing it could be is a crack in the side of the mountain. I need to get up there first thing in the morning and look at it."

Lucky was not sure the weather would be good enough for Cactus to fly one of the base's helicopters up there in the morning. Yunya was forecast to arrive at about 10:00 a.m. He called Cactus at the helicopter hangar and alerted them to be ready for a flight at 6:00 a.m. Then he folded his arms across his chest, bent over at the waist and peered at the drums. A few days earlier the needles had been moving sporadically, almost lazily, back and forth on the drums. Now, they scratched back and forth frantically, nearly banging off their stops as they slammed from side to side. Dark, angry lines scribbled on the graph paper.

NOTES TO CHAPTER 4

1 Fire and Mud, p 11.

2 Ibid., p 628.

3 Rand interview.

4 JTF-FIERY VIGIL SITREP, 121100Z Jun 91.

5 Interview, SSgt John M. Thomas, with author, October 26, 1998 [hereafter cited as Thomas interview].

6 Grime interview 1992

7 JTF-FIERY VIGIL SITREP, 121100 Z JUN 91.

8 Fire and Mud, p 81.

9 Ibid., p 10.

10 Philippine Flyer, special bulletin No. 1, June 13, 1991, p 2.

11 Ibid.

12 Fire and Mud, p 11.

13 Philippine Flyer, special bulletin No. 1, June 13, 1991, p.1.

14 Fire and Mud, p 11.

15 Message, AMEMB Manila to USIA, Washington DC, 170752Z Jun 91.

16 JTF-FIERY VIGIL SITREP, 121100Z JUN 91.

17 Fire and Mud, p 11.

18 Thomas interview.

19 Philippine Flyer, special bulletin No. 1, June 13, 1991, p 2.

20 Rand interview.

21 Philippine Flyer, special bulletin No. 2, June 14, 1991, p 2.

22 Author's journal, Chap 8, p 4.

23 Thomas interview.

24 Author's journal, Chap 8, p 4.

Chapter 5 BLACK SATURDAY

Stephen Spielberg, George Lucas, and Wes Craven could not have collaborated to produce a scene more incredible than the hours of June 15, 1991. Stephen King might have come up with something similar, but only on his best day and only with his most perverse juices flowing wildly.

As dawn slowly broke, those who were awake on Clark Air Base saw the clouds that had rolled in ahead of Typhoon Yunya, and they covered most of the Zambales Mountains west of the base. The lower slopes of Pinatubo were visible, but low scud masked the all-important summit. The weather station observers surmised that the low clouds around Pinatubo were temporary and would dissipate in the warming sun. They were right, and as the morning brightened the lower clouds moved away to reveal the mountaintop, now ugly and gray from the earlier eruptions.

Rick Hoblitt was awake and stuffing his backpack with items to take on his helicopter survey of Pinatubo. He was very eager to get a closer look at what he had analyzed as a fracture the night before. If it were a fracture, then the situation might be dangerous for Clark. All planning to this point had been predicated on a Mount Saint Helens type of blast being focused out to the northwest. A crack of the Clark side of the mountain was a different can of worms.

As Hoblitt nervously fussed around and packed, others wandered into the drum room and milled around looking for coffee or surviving Cheese Puffs. General Studer, Gator, and Colonel Grime, Eagle, leaned over the drums and looked at them and the mountain alternately.

The drums were saturated. The needles from every station swung steadily back and forth, "wall-to-wall" in volcano-speak. They were a solid black trace from edge to edge, an EKG measuring a thousand heartbeats per minute.

Lucky briefed Gator and Eagle on his intention to blow the siren at 6:00 a.m. and get everyone formed up at the gates. He had played a waiting game for the past few hours, and it seemed to everyone in the room that another eruption was imminent. Studer nodded his assent and went outside the building. Grime wanted reassurances that the helicopter crew knew that they were to take absolutely no risks in getting Hoblitt close to the mountain. The pilot for the morning was Capt Brent Nyander, who had flown many of the Pinatubo observation missions already. Nyander knew the terrain and exhibited a very healthy respect for the volcano. Grime agreed to the flight and turned away.

The Big One

Hoblitt was still puttering around the drums and chatting with the sleepy-eyed scientists. Just as the CAT commander walked over to remind him to get a move-on, the mountain blasted a huge eruption unlike any of the previous ones. It was 5:55 a.m.

The instant the eruption started, everyone, including the most novice volcanologists, knew it was "the big one." It was an enormous lateral blast, not a tall vertical one like the pre-

Just after dawn on June 15, 1991, Mount Pinatubo started its series of cataclysmic eruptions that were ten times larger than Mount Saint Helens.

vious "photo-ops," as Dave Harlow had called them. Almost instantly an area approximately 10 kilometers (6 miles) wide boiled up to the bases of the clouds. Lucky spoke urgently into his brick, "Hawkeye, Lucky, blow the siren."

Hoblitt turned to the CAT commander with a look of desperate accusation and shouted, "See that! You see that, goddammit! That's exactly what I didn't want to happen!"[1]

Col Bruce Freeman, vice commander of Thirteenth Air Force and JTF-Fiery Vigil* chief of staff, recalled the explosion. "The entire mountain ridge just opened up. I had a wide-angle [camera] lens from 12 miles away and it was wider than my lens."[2]

Mercury said, "I was staying at the command post and sleeping in my car . . . and I had parked my car so I was looking right at the mountain. Wherever you were, you were dead tired. I was sitting behind the wheel of the car, and I had gone to sleep in there about 4, and I woke up at almost 6, and the sirens were going off. Cars were all around me going just as fast as they could. I said, well, we must have gotten the order to evacuate. I looked up [at the volcano] and there was this wall going laterally instead of vertically. All the others that we were able to see were shots that went straight up. They just shot straight up until it was like they ran out of steam . . . . This one was just expanding as you watched it . . . didn't look like it was even 10,000 feet high and in the middle of it were lightning bolts. All I could think about was the geologists telling us the lateral ones are the killers."[3]

"All I could think about was the geologists telling us that the lateral ones are the killers"

Inside the Dau command post, some of the drums had already stopped working. The volcanologists assumed that the blast destroyed the stations nearest the mountain. The ash spread rapidly across the horizon and huge sheets of pyroclastic flows cascaded down the flanks of the mountain. Gator, Eagle and Lucky huddled for a moment in the command post and quickly made the decision to abandon the base.

Lucky told Hawkeye to instruct everyone to continue the evacuation to the college. The beards and the CAT grabbed their knapsacks, backpacks, and gym bags and hurriedly left the command post.

One principle of an emergency evacuation is to check to make sure everyone knows an evacuation is taking place, so Gator and others swept through the building looking for stragglers. At the back of the building, he discovered Personnel Bobbitt pitching cases of MREs into a truck.

Colonel Freeman recalled the story. "General Studer ordered him [to get moving]. As Bobbitt continued to throw the cases into the truck, he said, 'I can't leave yet, General, I've got to take care of my guys.' The general insisted. Bobbitt threw another armload of food into the truck and dashed off for the college."[4]

As Lucky swept the building he found Mercury on the telephone with a reporter from *Stars and Stripes*, who had not been particularly friendly to the base in her reporting over the past couple of years.

Lucky said, "C'mon, Merc, we gotta go."

Rand smiled and laid the phone on the table. As they strode from the room, both could hear the reporter saying, "Ron? Ron? Are you still there?"

Long strokes of lightning flashed through the ash cloud, the byproduct of enormous static charges from the friction of the massive explosion of ash into the air. In just a matter of a few moments, the first feeder band† from Yunya had moved across the base and a light rain started to fall from the darkening clouds.

Lucky took two of the volcanologists in his staff car, and they joined the convoy on the MacArthur Highway just outside the Dau gate. The locals seemed unperturbed by the stream of vehicles joining their usual Saturday morning traffic. The convoy was moving slowly, and an hour later the car had traveled only half the distance to the college.

Ed Wolfe, who had replaced Dave Harlow only a few days earlier,‡ admitted he had second thoughts about leaving. He reasoned that some of the seismographs were still operational, so it was still possible to monitor the volcano. Further, he stated, any PFs that had come from the blast would be gone by now. Lucky asked him if he thought they could return, and after some discussion among the geologists, Wolfe said he was willing to go back. In fact, he wanted to go back.

The feeder band from the typhoon had already passed and the sun was shining. Lucky pulled the staff car off the road into a dry rice paddy and started working the radios with Gator and Eagle. He explained the situation. They

* *Joint Task Force-Fiery Vigil was activated June 10, 1991. It is discussed in greater detail in Chapter 7.*

† *A thin band of clouds and rain preceding the main body of the typhoon.*

‡ *Harlow was en route to the United States as part of the USGS scheduled rotation of personnel.*

quickly formulated a plan to take two 44-man security police flights, the volcanologists, and a few others back. Their objectives would be to monitor the volcano and guard the weapons stored in the MSA. In fact, neither Gator nor Eagle had left the base. Both were waiting at the main gate, reluctant to leave the helm of their command until they were sure everyone was gone.

The small group of less than a hundred formed up in the dry rice paddy and headed back to Clark. When they drove back onto the base, they found it just as they had left it. No ash. Everything was green. Clouds again shrouded Pinatubo.

Although Jim Weed and his workers had switched off the main power station, the generators at Dau were still running. The air conditioning was still on, and as the small band entered the drum room, nothing had changed since they had left. It appeared that their hopes were realized, and the prevailing winds were carrying the ash away from the base.

"I had a wide-angle [camera] lens from 12 miles away and it was wider than my lens"

The security police quickly re-established security around the MSA, and Hawkeye controllers checked out communications gear. Volcanologists returned the few drums still alive to reporting status. Everything worked properly, powered by the dual-redundant emergency generators built into the old building.

Gator and Eagle conversed with people at the college via the bricks. It quickly became evident to everyone that the brick net's central station at Dau provided communication to all the Ash Warriors. The system was not designed to work at such long distances, but it was doing the job. Lt Col Charlie Hall's communications troops had done a superb job hooking everything together.

Some security forces not on post at the MSA resumed patrolling the housing areas and some storage areas that contained valuable items. Their Philippine Air Force law enforcement partners had stayed on the base. A few looters were spotted and chased away, but there did not seem to be many thieves waiting for their chance to rob an American home or warehouse.

Jim Mori, a newcomer to the USGS-PHIVOLCS team, had arrived that morning from Hawaii to study the volcano. He wrote, "I had left the hotel in Manila and had a leisurely drive to Clark Air Base with an Embassy driver. When I got to Clark, the USGS people were milling around the front gate waiting to get back on the base. They looked terrible and didn't seem to have a very positive attitude about things in general. I was beginning to wonder what kind of situation I was getting into. Driving into PVO that morning was the only time I saw anything green on the base.

"One of the PHIVOLCS people brought doughnuts into the observatory [drum room]. Everyone seemed to really enjoy that. I only realized the significance of that later after 2

Different from the preceding eruptions this was a tremendous lateral blast approximately 6 miles wide and quickly reaching the base of the clouds.

weeks of MREs. (Which reminds me of the scrambles when a new box of MREs was opened, and everyone tried to get their favorite meal.)"[5]

Lucky, who had not left the command post for nearly four days, decided to get a view of the base from where the security forces were working. He drove from Dau westward across the runways and the parking apron, then went up Dau Avenue past Chambers Hall. He turned left at the high-rise, then right on Wirt-Davis and slowly cruised west along the parade field. Before he got to General Studer's quarters at the far west end of the parade field he heard Gladiator telling Eagle on the brick that it was getting dark with ash on the hill. Gladiator said he was withdrawing his troops from the area because the ash was too thick to see anything anyway. Eagle agreed.

The CAT commander stopped his car on Wirt-Davis and waited. Suddenly, the area behind Thirteenth Air Force headquarters started turning black from the ground to the sky. The darkness was moving very slowly east and as it moved past street lights and porch lights, those lights disappeared as though they were switched off.

Pumice rocks, the size of peach pits, were pelting down among the raining mud, and one earthquake after another slammed across the base

The CAT commander transmitted on his brick, "Eagle, Lucky."

The wing commander responded immediately, "Go ahead Lucky."

"The ash up here is very thick. Get everybody back to Dau right away."

"Roger that, Lucky," Colonel Grime answered, and the brick net was suddenly saturated with calls between Hawkeye, Eagle, Gladiator, Pioneer, and other officers out with the police to withdraw to Dau.

Clouds of volcanic ash rained down on the base and for many miles into the surrounding countryside. Smaller eruptions would have the same result for weeks afterward.

The black ash cloud was only a hundred yards from Lucky's staff car when he did a 180 turn over the curb, through some shrubbery, back over the curb onto Wirt-Davis. He turned on his emergency flashing lights, for no good reason since there was no one to see them, and raced the ash cloud east towards the Dau command post.

By the time he got to the flightline, the ash cloud was almost upon him, and it was starting to rain again. The darkness chasing him was like watching the old television soap opera lead-in for "The Edge of Night," where the darkness of night moved slowly across the screen. There was daylight in front of him, but right behind him was a black curtain. When he was only a hundred yards from the Dau command post, the cloud enveloped the staff car. He stopped. Forward visibility was less than ten feet, and the soft ash, mixed with light rain, covered the windshield. He turned on the windshield wipers, but they only made a chocolate mess of the stuff. The windshield washers took enough of it away to keep the car creeping slowly forward, but he had to hold the washer button down constantly for the washers to be effective. He knew he had arrived when the front of the car bumped into something solid – the corner of the Dau command post building. He turned off the car, jumped out, followed the wall to a doorway and stepped into the building.

The drums, the weather radar and the weather station's barograph told the story. At 10:27 a.m. the volcano went into a series of continuous explosive eruptions. The weather station's barograph, which measured atmospheric pressure in inches of mercury, pulsed each time the volcano erupted.[6] The volcanologists used the pulses to help determine the start of an eruption when it was not possible to see the volcano. Some base personnel, most notably General Studer, could feel the atmospheric pulse in their ears whenever the volcano exploded.

Observers in PVO watched the ash move across the runway. The emergency lighting along the airfield had not been turned off in the rush to evacuate right after sunrise. One by one the lights disappeared as the ash cloud enveloped them. By midday it was totally dark outside.

Hunkered Down

Winds from Typhoon Yunya weakened when it made landfall on the southeast shore of Luzon, and the typhoon became Tropical Storm Yunya. However, the storm continued to drop large amounts of rain as it sauntered northwest and passed about 50 miles from Pinatubo.

Typhoon rain turned to cement when mixed with falling ash.

Eagle and Gladiator withdrew all security forces to the Dau command post, and there they waited. Many stood under the building overhang and wondered if it were raining mud or mudding rain. Satellite photography would later reveal that the top of the ash cloud was nearly 240 miles in diameter at an altitude of 18 miles.[7]

Shortly after the ash cloud moved over the base, the earthquakes started, and these were earthquakes that pounded through the base and rattled the command post. The volcanologists were able to estimate the magnitude of the temblors, and most ranged between 3 and 5.6 on the Richter scale.[8] A discussion began among the volcanologists and the officers about the possibility of a caldera forming. As ash and other materials were blown from a volcano, a chamber would start to form under the mountain. When the roof of the chamber became weak enough, it would collapse into the void and, in turn, be blown out of the volcano by the eruptions, which, in turn, would cause the void to enlarge, weaken, and then collapse again. This cycle would continue until the volcano ran out of energy. If the eruptions were big enough, the end result would be a crater where once was a mountain. The scientists agreed that Pinatubo would become a caldera before it was done. The only question in their minds was how big the caldera would be.

At 2:00 p.m., only one monitoring station was transmitting data back to PVO; the other six had been destroyed by the volcano. Pumice rocks, the size of peach pits, were pelting down among the raining mud, and one earthquake after another slammed across the base.[9] Those outside could hear a continuous low roaring noise that some equated to the sound of a subway train in the distance. The volcanologists identified the eerie roar as lahars (mudflows) raging through the Abacan and Sacobia rivers on either side of the base.

One reason for selecting the Dau command post as the alternate site was that the route from Dau to the Pampanga Agricultural College crossed no bridges. Eagle sent a scout team to check the route to make sure it was passable. While they waited, they ordered everyone into the center room of the command post. It seemed the logical action to put as much building between them and a mudflow, or even a small pyroclastic flow.

The new volcanologist, Jim Mori, said, "I was scrounging around the seismic gear, trying to look at the data and recording system when Tom Murray told me that everyone was moving to the back of the building. I recall walking back there with everyone and then hearing about the possibility of pyroclastic flows coming our way. To me it was a very surrealistic scene with everyone seemingly calm in the midst of all the stuff coming down on top of us from the volcano. For some strange reason I was very conscious of standing on the false floors in that room. I was thinking, 'What in the hell was I doing in this place? This was sure a stupid idea, volunteering to come to the Philippines!' I had spent the night in a nice hotel and had a good breakfast several hours before in Manila, and now suddenly I was standing in this strange stark room with a bunch of people I didn't know, waiting for our building to be overwhelmed by a pyroclastic flow."[10]

A hot lahar roars down the Abacan River watershed very close to Clark AB.

The Final Bugout

The scout team reported that the evacuation route to the college was passable, and Eagle ordered the small team to abandon the base. It was no longer possible to guard anything, and the only remaining objective still attainable was to get everyone out of harm's way.

The building emptied immediately as people grabbed anything they could carry and headed for their vehicles. Rand's account of the final evacuation tells the story vividly.

"Everyone was exhausted. There's an adrenaline rush that keeps you going, but in a lot of ways you're brain dead . . . you're just not making good decisions . . . you do stuff instinctively or automatically that turns out to be right, but you don't think it through. Maybe that's a function of experience and training and you're just responding that way. I know everyone was real tired, the geologists, the commanders, everybody was just watching and waiting and seeing what it was going to do. I know I thought we'd made it through and nothing was going to be happening, [but] about noon it turned pitch black. I mean it was so black that everyone went outside. It was darker than any night I've ever seen. And it was the first wave of all that ash that we'd seen in the morning, the first wave of it finally reaching us. But then it just came down in a torrent and it was mixed with rain from Typhoon Yunya, and it was just unbelievable. As you stood out in it, it began to get chunky, it wasn't just drops of mud anymore, it was little pebbles and it started out kind of grainy, then it got bigger and more granulated and then it got [to] golf-ball-size stones coming down.

"Around 2:00 p.m., the building we were in began just shaking, really badly. They lost the last of the instruments they had on the mountain, and that's when they said now is the time for all the good guys to leave Clark. These [earthquakes] were some pretty bad ones, and . . . the building was really shaking so badly that I was beginning to edge near the doorway because I wanted to be able to get out quickly if it started to collapse.

"They said let's go. I still didn't have a map, I still didn't have the directions to the college. Visibility now was a hundred times worse or a thousand times worse than it would have been in the morning, but I figured that's not a problem. I'd just get in this caravan of vehicles and follow them up. You couldn't see that far in front of your car, even when you had your windshield wiper washing because of the darkness and the rain and the ash fall. You couldn't see the lights of the car in front of you.

"I remember carrying out my case of ginger ale* and I grabbed a sleeping bag and all the notes I'd been making for the last week or so and a case of MREs and I went to the car with all this stuff pelting me and everyone running for their cars. I said, good, I'll just fall in line behind someone here. By the time I put the stuff in the car, got the engine started, and drove to where I knew the road was, there weren't any other cars in the line.

Those from the north were fleeing south, and those from the south were fleeing north

"Well, I knew how to get off base and out to the MacArthur Highway and to the freeway, but after that I had no clue where we were going. Couple of minutes after I got in the car the windshield washer stopped working because the jets got all clogged up from the mud, so I had to roll my window down and start using the ginger ale. I'd just hold the can at the top of the windshield where I was looking and just have a steady slow pour with the wipers going and keep the little sliver of the windshield open. But it was really rocky, muddy stuff that was coming down and the wiper couldn't do a very good job.

"[There was every] manner of pedestrian that you can imagine, bikers, people pushing carts with all their belongings on it, people riding in carts being pulled by water buffaloes. [It] looked like everybody in the city was evacuating to the same place we were, the same direction we were going.

"Once I got out onto the freeway, [I] was completely lost, and I was hesitant about using my brick. I had never learned how to use the private function on the brick. There was a way you could dial somebody's brick number and talk only to him without talking on the open net. I hadn't learned how to do that, and I didn't want to get on the brick with maybe hundreds of people listening in and say, 'This is Mercury and I'm lost.' Anyhow, eventually, I did that.

"I called Col Art Corwin, who was the [deputy] security police commander, his call sign was Pioneer. I said 'Hi, Pioneer, this is Mercury . . . I think I'm on the National Highway and I don't know for sure where we're going and wonder if you could give me some directions.' I knew Pioneer was out there somewhere because he was one of the last ones to leave the base.

* *The CAT commander had directed everybody to take a case of soda to supplement the limited drinking water at the college.*

Base housing after the cataclysmic eruptions. Some called it "nuclear winter."

"Immediately, I mean even before he could respond, General Studer got on [the brick] and said, 'Where you at, Merc?' and he gave me directions. Well, you couldn't see where you were going anyhow, so I just decided to follow the traffic and see where it went. I finally got to the college about 18 or 19 cans of soda and 4 hours later to go the 10 miles."[11]

Many in the convoy to the college shared Colonel Rand's experience. The streets in Dau and Mabalacat were choked with terrified Filipinos who were fleeing in every direction. One Mabalacat city official was standing in the middle of the street trying to unsnarl a melange of jeepneys, bicycles, mopeds, and wooden carts loaded with refugees. The total chaos was beyond control. Those from the north were fleeing south, and those from the south were fleeing north. Thousands only wanted to go in the same direction as the Americans.[12]

Jim Mori left the Dau command post with his US Embassy driver. "As people went outside the building they began to frantically throw things in the trucks and cars and headed out as fast as they could. The one exception was that great guy in charge of supplies [Personnel Bobbitt], who was handing out the six-packs of Cherry Coke at the doorway.

"My most vivid memories of that day are stepping outside the Dau complex into the mess coming down. It was pitch black outside and we couldn't see more than a couple hundred feet. It was pretty disconcerting to have all this sandy material coming down out of the sky, along with being pelted by the bigger chunks of pumice. The sounds were overwhelming with the clatter of the pumice hitting the metal roof and cars and the thunder going on above us. One of the most frightening things was this low roar in the background that . . . probably was a lahar going down a nearby river valley. I was pretty staggered when I first got outside, but after I regained a little composure, it was interesting watching other people come out the door. I could tell from their expressions they were experiencing the same sort of feelings of amazement.

"The entire drive out to the agricultural college, I was trying to keep my driver calm. He kept saying things I couldn't understand and crossing himself. I'm sure I was just as scared as [he], but I figured we would never get to our fall-back position unless I showed some confidence. We would have to stop every 10 to 15 minutes to clean off the windshield and we independently came to the same conclusion as all the other vehicles that Cherry Coke made good windshield cleaner fluid. When I was packing for my trip, my wife had given me a pair of swimming goggles, saying that it might help me to see during an eruption. I thought it was a little silly at the time, but it turned out to be a real lifesaver. Taking my turn to go outside to clean the windshield, it was hard to see and painful on the eyes, without wearing the goggles."[13]

When the evacuation order was given, Lucky ran through the command post building searching for sleepers that had not heard the order. Once assured the building was clear, he grabbed a case of ginger ale and went out to his staff car. It was covered with two or three inches of heavy, wet ash. He pushed most of the sludge off the driver-side windshield, threw the case of ginger ale on the passenger seat and started the car. When he tried the wipers they worked, but the windshield washer was empty. He had forgotten to refill it after his mad dash back to the command post earlier in the day. He looked around and could see no other vehicles, and just as he was about to start walking a large, security police, four-wheel-drive vehicle came around the corner and stopped beside him.

He opened the passenger door of the jeep and looked inside. A young airman was driving the vehicle. In the rear seat was a young man

Lahars washed across the main road to Hill Housing.

Lucky recognized as one of General Studer's bodyguards. On the seat beside the bodyguard sat a small dog that was whimpering and shivering as he cowered against the bodyguard. The young driver looked at Lucky and said very calmly and firmly, "Colonel, if you're coming with us, you need to get in. Right NOW!"[14]

As the falling pumice stone tattooed the roof of the vehicle, another earthquake slammed through the parking lot. The CAT commander retrieved his bugout gym bag from the staff car, threw the case of ginger ale on the floor of the jeep and jumped in. The driver started moving slowly forward, found the exit gate and turned toward the main highway.

As the jeep moved slowly into the seething mass of refugees, a young woman walked beside the car and held up her infant child. "Take my baby, please. Please!" she pleaded. Lucky rolled down his window and shouted, "No. No. Keep it with you; we'll never be able to find you again. You'll be okay. Keep walking that way," he pointed toward the college. The young woman, tears streaming down her face, melted into the crowd. She vanished in the throng of people that carried clothes on their heads and small children in their arms.

"Colonel, if you're coming with us, you need to get in. Right NOW!"

Hell at Subic Bay

At Subic Bay, Cubi Point, and San Miguel naval stations, where 15,000 Air Force evacuees and several thousand Navy troops and families were crammed into every nook and cranny of the three installations, the situation was no better and continued to get worse throughout the night. The ash fall and earthquakes were just as severe for them as for the Ash Warriors huddled at the college.

The inlet of Subic Bay was the Navy's best deep-water port in the western Pacific. Subic Bay Naval Station was the main base of the complex. Cubi Point Naval Air Station abutted the main base and featured a runway capable of accommodating large commercial and military transport aircraft. Approximately 15 miles northwest of Subic Bay sat San Miguel Naval Station, a small, former Vietnam-era communications facility.[15]

Subic Bay and San Miguel were 20 miles from the volcano. They were selected as evacuation locations because USGS-PHIVOLCS advisors and Clark's planners thought the Navy complex was well outside the reach of deadly pyroclastic flows. They also thought the Navy bases would be clear of heavy ash fall. In this second thought, they were wrong, for ash and mud poured down on the bases in the same quantities as Clark, which was only half the distance from the exploding mountain. The wild card that trumped the planner's good intentions was the inopportune arrival of Typhoon Yunya.

In May, USGS-PHIVOLCS scientists, in concert with Air Force meteorologists, produced an ash fall warning map. The map predicted where ash might fall, and the predictions were based on the seasonal, prevailing winds. The areas predicted for exposure to moderate ash did not include the Subic Bay complex.[16]

As Yunya marched through central Luzon, its winds encountered the enormous ash cloud hanging over the bases and the effect was like a whirling electric mixer being thrust into a bowl of pancake batter. Instead of being carried away by prevailing winds, the ash was continually recycled over the bases and, soaked by Yunya's rainfall, it clumped to the ground in prodigious quantities.

Clark evacuees and Navy residents cowered in shelters and homes as earthquakes rolled through the base, huge trees crashed to the ground, and buildings collapsed throughout the afternoon and night of Black Saturday. As at Clark, the raining goo blocked out the sun. Electrical power had been turned off the previous day and total darkness enveloped the Navy complex, punctuated frequently by colored lightning bolts that were blue, red, or green.

One evacuee, Eagle's wife and the mother of their two young sons, wrote in a letter to family and friends, "The sky is very dim, and getting darker by the hour. The typhoon, now a tropical storm, is upon us, with the afternoon and evening to be the worst. Somebody discovers that the Officers' Club is serving coffee and some food. The ash and rain continue to fall. Again, we sit outside and listen to the radio. Apparently the big eruption has happened/is ongoing. [We] take our umbrellas and walk a half block to the club for lunch, where they have a sandwich bar set up. There are lots of candles burning and a pianist plays in the background. At noon, I notice that, looking out the window, you can no longer differentiate between land and sky – it is totally pitch-black. You can't move around outside at all without a flashlight.

"The rest of the afternoon is spent at our room, out on the front 'porch,' looking out at the blackness, listening to the wind and rain. Soon we are listening to loud cracks as the branches from the huge old trees that surround our building begin to break off with great crashing noises. We also begin to feel the tremors, some of which are quite strong indeed, which go on more or less constantly all day and night. They don't bother me much because we are essentially outside already in case things start falling, plus last summer's earthquake registered nearly 8 on the Richter scale, and Clark suffered little from it. What bothered me more was the unrelenting barrage of thunder and lightning, all of which was right on top of our heads from the sounds of it. I was really sick and tired of that before long. The boys stayed occupied playing cards and other games by candle light, with two . . . neighbors at Clark, whose mother was nearly smushed [sic] by a huge branch that fell up onto the porch, where she was sitting, stopping about 6 inches from her feet.

"At some point, the local military radio station went off the air, and our water quit running, so things were getting pretty grim. Our roof started making cracking noises periodically, so we'd frantically search with our flashlights for signs of collapse, but could never find anything. The weight of the wet ash was tremendous, and one section of gutter in the front where we sat broke off, so we kept expecting the worst. Finally, we [would] drag off to bed reluctantly – it's so hot in the rooms, plus you have to burn a precious candle to see anything. Lying on the bed, sweating, shaking with the tremors and worrying about the roof falling in is not conducive to a good night's sleep, believe me.

"Sure enough, about 5 a.m., there's a great loud CRACK! I leap out of the bed, shout 'Get out!' at the boys, smack into Ricky who has jumped up in the pitch black, Stuart is shouting for the flashlight so he can get down from his (very high) top bunk. We all get outside with everyone else, to discover the overhang all along the back side of the building has fallen down. The good news is that we realize the sky is not black anymore – just dim gray, but still daylight."[17]

Navy engineers, augmented by about 120 of Col Jim Goodman's engineers from Clark, moved frantically from building to building at the Navy bases in a effort to assess the structural integrity of buildings loaded down with wet ash. Goodman recalled, "There was a building right across the street from the Sampaguita Club that we were able to get about 130 people out of about a half-hour before it caved in."[18] Rescuers guided the frightened mass into the Sampaguita Club where they packed into the building with others seeking refuge from the ash fall. During the rest of the night, officers and sergeants tried to keep people calm in hot, close conditions with no light and the threat of the building's concrete ceiling caving in at any minute. To abate the threat, Goodman assembled a team of volunteers to go onto the roof to clear away the five inches of wet ash that had accumulated. Not many tools were available, and some of the volunteers shoveled with plastic cafeteria trays.[19]

Others were not so fortunate. When the roof of the high school gymnasium started to creak and groan under the weight, sentries hurriedly ordered everyone from the building and evacuees scrambled for their lives. Nearly everyone made it, but not all. Falling debris killed the nine year-old daughter of an Air Force couple and their eighteen-year old niece.[20]

The College

The ash did not fall as heavily at the college. The streets of the small campus were jammed with vehicles of every size and description. Three buildings in the center of the campus formed a large U-shape. Vehicles packed the area inside the U. Exhausted troops filled all the classrooms on the ground floors. Many had simply dropped everything where there was an open space and then collapsed on top of their gear. The school gymnasium and auditorium was wall-to-wall troops. Conversation was muted as most sat quietly and contemplated the previous five days.

General Studer and his JTF-Fiery Vigil staff, along with the CAT, reconstituted their respective headquarters in a large classroom on the second floor of one of the larger buildings.[21] The mood in the room was somber as communications troops hooked up radios to portable generators.* A quick accounting of personnel by unit had started as soon as the first of the mission essential team arrived at the college. When the last 100 arrived late that afternoon, the count was complete and all were accounted for.

Several people carried small commercial radios, and one of them reported that news bureaus in the United States were saying that everyone at Clark had been killed. Colonel Rand overheard one of the reports and found a combat communications team with their portable equipment in a jeep. The clever communicators were able to contact a station in Manila that patched him to Guam that connected them to a station in Hawaii that connected them to a station in Atlanta that connected them to Cable News Network (CNN) headquarters. The effort took over two hours, but when it was complete, Mercury was able to sit on a wooden stool beside a jeep in the middle of the Philippines and report directly to CNN. He told them that all

** One high-frequency (HF) radio and two ultra-high frequency satellite communication (UHF-SATCOM) radios.*

1,500 of the mission essential force were safe at the Pampanga Agricultural College. One person had a broken arm, but there did not seem to be any other injuries.[22]

Colonels Murphy and White had planned well. Everyone that wanted an MRE had them, and several water trailers were set up at strategic points to serve the exhausted men and women. The thousands of refugees that followed the Americans to the college were a thorny problem. The college was bursting at the seams with the Americans, and the Filipino refugees could not go any farther. So, they stopped and slept where they were. Compassionate airmen "commandeered" some water and rations for the refugees nearest the college, but the masses could not be reached and thus spent a miserable night on the flanks of Mount Arayat, the extinct volcano ten miles east of Clark.

Although the ash fall subsided considerably, rain from Yunya continued to soak those not under cover. The frequency of the earthquakes reduced, but the ones that came were hard jolts that rattled the buildings.

Once plans were made to return to the base at first light the next morning, the mission essential force quietly settled in for the night. Many refused to sleep in the college buildings because of the constant earthquakes and slept in their cars. Lucky, and the security police commander, Gladiator, slept in the back of Gladiator's command vehicle. They had to put the rear gate down to stretch out, so throughout the night, Yunya's rain soaked their sleeping bags. However, as the earthquakes rolled through, the vehicle rocked gently, and they slept the sleep of the dead.

Mercury recalled, "The guys we were with showed me the room they had staked out some floor space in. I was only in there 20 minutes. It was on the second floor of this concrete building, and I said I didn't live through the evacuation to get crushed in this building tonight. My NCOs* and I found a little *sari-sari*† store . . . which was selling warm San Miguel‡ beer. We had two or three beers and I went back to my car and was asleep before my head hit the back."[23]

Harlan E. Mikkelson, then a staff sergeant in the mounted horse patrol, recalled. "When we went out to the college that first night, we slept in an auditorium. Everyone was sleeping on the floors with whatever they had, and an earthquake shook the building, and no one even moved. The earth was shaking, and no one moved. It was like, 'I'm too tired to move.' No one got out. No one runs to the doors. No one got excited or nothing. It was like, 'I've been through too much already.'"[24]

The volcanologists gathered in a room down the hall from the makeshift command post. Mori recalled, "Sitting there in the dark in a second floor room of one of the buildings, we were feeling lots of earthquake. There was not much to do so we began to time the events, which were coming at the rate of more than one a minute for a couple of hours. I recall Rick Hoblitt pulling out one of [his] books and, using a little penlight, [beginning] to read aloud. He read accounts of what was thought to have happened at the caldera-forming eruption at Crater Lake, and we started talking about what should be the appropriate fallback distance for our situation."[25]

Whether they were far enough away was a moot point. Lying on concrete floors, sitting in car seats, curled up on field packs, unable to even brush the caked ash from hair and clothing, the exhausted force of Ash Warriors could do no more.

** Non-commissioned officers.*

† Primitive convenience store.

‡ Local beer favored by generations of military men and women serving in the Philippines.

NOTES TO CHAPTER 5

1 Author's journal, Chap 9, p 1.
2 Freeman interview.
3 Rand interview.
4 Freeman interview.
5 Letter, Jim Mori to author, n.d. [hereafter cited as Mori letter].
6 Fire and Mud, p 627.
7 Ibid., p 13.
8 Ibid., p 630.
9 Ibid., p 13.
10 Mori letter.
11 Rand interview.
12 Interview, Guy Hilbero with author, October 30, 1991.
13 Mori letter.
14 Author's journal, Chap 9, p 4.
15 Fire and Mud, p 77.
16 Ibid.
17 Letter from Julie Grime, July 1991, on file PACAF/HO.
18 Goodman interview.
19 Letter, Goodman to family and friends, June 1991.
20 US Embassy SITREP, 160857Z Jun 91.
21 Letter, 13 AF/CC to USCINCPAC, Subj: First Impressions Report-Operation Fiery Vigil, July 16, 1991, p 1.
22 Rand interview.
23 Ibid.
24 Mikkelson interview.
25 Mori letter.

Chapter 6 NUCLEAR WINTER

On the morning of June 16, 1991, the Ash Warriors, led by Col Bill Dassler, cautiously returned to Clark Air Base in two humvees* and a Jeep Cherokee. Crowds of refugees from Angeles, Dau, and Mabalacat lined the roads as the small convoy eased through the masses. The crowds silently watched the small American contingent pass.

The day was hazy, almost like a bad smog day in the Los Angeles basin, but the reduced visibility was from small flakes of dry ash in the air. The closer to Clark the convoy came, the deeper the ash. Collapsed buildings littered the villages through which they passed. Roofs, trees, vehicles, sidewalks, streets, and signs were caked with a thick, dark gray layer that was the result of the collision of Pinatubo and Yunya.

Dassler and his men, who thought they were the first security forces to return to the base, got a surprise just outside the entrance to the base. Security police forces of the "Town Patrol,"† were on the MacArthur Highway directing traffic. When Dassler stopped to inquire as to how and why they had returned so quickly, he was told by a young sergeant that they really had never left. Their job was to direct traffic during the evacuation, so they just stayed and directed traffic rather than abandon their posts.[1] Their uniforms nearly were unrecognizable under the coating of ash. A thin powder covered their skin where it was bare. To Dassler, it seemed amazing, but they were still wearing the blue beret that is the trademark of USAF security forces. Somehow, they had managed to keep the caps relatively clean.[2]

Taking Stock

The small convoy entered the base near a cemetery that held several hundred remains of American and Philippine military veterans. The American flag still flew over the grave markers jutting out of the ash, a red, white, and blue splash of color over a dead-gray sea.

Gladiator directed one vehicle to check the MSA and the other to check the main armory where hundreds of weapons were stored. The emergency generators at the Dau command post were still chugging away, thanks to the elaborate air filtration system that kept ash out of the motors. The generators powered the brick system, so Dassler was able to radio from his jeep back to the college to send more security forces. He answered everyone's first question when he said the base was still there. The worst of the ash fall seemed to have passed.

Eagle, implemented a phased approach to restoring operations at the devastated base. Phase 1 initiated a small reconnaissance team to assess the damage. Phase 2 included USGS scientists and a small command element. Phase 3 brought in a 44-man security police flight to secure the munitions storage area. Phase 4 initiated plans to assist Filipino refugees.[3]

A second contingent of security forces, as well as General Studer and the wing commander, was not far behind Dassler and entered the base a short time later. As they maneuvered around debris, they were stunned by what 12 hours of ash, rain, and earthquakes had made of the once-beautiful base.

** High mobility multipurpose wheeled vehicles.*

† The Town Patrol was a special unit of the security police that patrolled the liberty areas surrounding Clark. It was composed of law enforcement officers of the USAF police, Philippine Air Force police, local police and the Philippines Integrated National Police.

The convoy returning to Clark AB paused temporarily as a farmer surveys his devastated fields.

Eagle, the wing commander, recalled, "We came to the main gate, and without sounding too melodramatic, it almost brought tears to the eyes. We had a base we were really proud of . . . we'd put a lot of time and effort into making it the best it had ever been, and you get back and you see virtually all of the trees destroyed . . . everything covered with ash . . . the buildings collapsed.[4]

Some thought it looked as though a nuclear war had been fought there. Others thought it looked like a terrible blizzard had swept across the province, while others thought it looked like the moon. Everything was dark gray. Six to nine inches of heavily compacted ash covered everything. The streets and sidewalks were indistinguishable. Gray covered even the street signs. There was no color on the base, only a sea of gray. Most eerie of all, though, was the silence. One could stand in the middle of the base and not hear a sound. It took a moment to realize why it was so quiet – all the birds were gone, too.

Many Ash Warriors, after they returned to the base, related the same sense of total disorientation they felt when they saw the base. One could stand in an area that seemed familiar, but at the same time feel totally lost. The huge acacia trees, once proud and beautiful, lay in gray heaps. All the other trees, shrubs, and bushes were covered or flattened. Roofs were covered and even the walls were hidden behind a thin covering. No matter which way one turned, everything looked the same – dark gray. It was disorienting to stand some place that felt familiar, yet have no idea which way to turn to go to a desired destination. The base was a lumpy, gray blanket as far as the eye could see.

The wing commander immediately put some engineers to work assessing the damage, but much of the devastation was obvious. What had once been fighter jet shelters along the flightline were flattened from the tremendous stresses of wet ash and earthquakes. Several large supply warehouses, which housed hundreds of millions of dollars of equipment on some 320,000 square feet of floor space, had caved in. The base gymnasium looked like a crumpled piece of wastepaper thrown carelessly on the floor by a giant hand.

Buildings and homes covered with ash from Mount Pinatubo.

Personnel return through the same gate that they had left.

One of the hangars, which was designated as a sleeping area and rally point to replace the tent city, had collapsed. It was the one the CAT selected to provide shelter from Yunya's winds and rain. Dumb luck had intervened, however, because the hangar was never used. The evacuation to the college had saved many lives. Crushed cots and bedding jutted and fluttered out of the tangled mass of ash and steel that had once been the hangar roof.

Most streets were impassable because a jungle of broken tree branches littered nearly every street. Some areas were accessible by four-wheel-drive jeeps and the larger, more powerful humvees. It was nearly impossible to get into many of the housing areas, even on foot. Countless downed power lines laced through the broken branches.

The USGS-PHIVOLCS team found the PVO to be next to useless. Only the seismometer at Clark was operating; the other six stations were buried under ash or destroyed by the massive eruptions. They were unable to determine the scale of the eruptions but were confident it was of "historic" proportions.[5] They estimated that the eruptions were at least as big as the Mount Saint Helens eruptions.

Thick ash in the air shrouded Pinatubo, and although scientists wanted to see what was going on, it was impossible to fly them up there. The ash was still thick in the air, and occasionally a large black mass of it would drift across the base. Based on information from satellite photographs, which the volcanologists received on a cellular phone, they learned that the top of Mount Pinatubo was no longer there. In its place was a crater 2 to 3 kilometers (1.2 to 1.8 miles) in diameter.[6]

Pyroclastic flows filled the pre-eruption topography across the Zambales Mountains. Valleys, which were once hundreds of feet deep, were filled. Crow Valley, which had been the site of the bombing targets used during Cope Thunder exercises, was nearly filled with pyroclastic flow.

A few, who were able to find a path through the debris into the Hill housing area, saw that one pyroclastic surge had stopped in the river bed only 400 yards from what had once been the homes of Air Force families. The PF had stopped almost exactly where the beards' warning maps had predicted.[7]

The scientists opined that the worst eruptions were over, but they were not sure. Further, they assumed eruptions would continue but not on the huge scale seen before. They envisioned a scenario in which the eruptions would be more frequent but not as powerful. They estimated a period of six months to two years of continuing eruptive activity.[8] Hoblitt and Wolfe believed the volcano would have to be closely monitored for the next three to five years.

The base was a lumpy gray blanket as far as the eye could see

Lahars

The volcanologists were most concerned about the likelihood of lahars. Before the eruptions started, they instructed the CAT and the commanders about the dangers in sessions that became known as "Mudology 101." Now the lessons were becoming a reality. The volcano had covered the area with uncountable millions of tons of ash. When it got wet, the ash would absorb the rain water until it became so heavy that huge chunks of it would scab off. These enormous flows of wet ash, the consistency of wet cement, would then flow wherever gravity would take them. If there was a large amount of rainfall, the lahars could sweep away everything in their path including stones and even large boulders.

The scientists were certain the lahars would occur since the rainy season was imminent. Considering the amount of ash Pinatubo had ejected and the heavy monsoon rains that usually started in June or July, there was no doubt in their minds that the provinces around Clark, especially in low-lying areas, would experience the devastation of lahars for many years to come.

At this point, the Ash Warriors had already gotten a taste of what lahars could do. After checking out the riverbeds it was clear that the roaring noise all had heard on Black Saturday was the rumbling of lahars roaring down the rivers.

Several smaller lahars had washed through the base as well. Over the many years of Clark's existence, an intricate system of drainage ditches were built throughout the base. The ditches were 10 to 15 feet wide and 6 to 8 feet deep. During the rainy season, these ditches carried the water from torrential rains through the base and out into the watersheds of the Abacan and Sacobia rivers to the north and south. However, Yunya and Pinatubo had filled the ditches to level with ash.

The eruptions of Black Saturday, combined with Yunya's rains, totally disabled the system. As ash and rain poured down, the rain, with no where to go except where gravity sent it, flowed across the base in enormously powerful sheets. One flood had slammed with such force into the enlisted club that it blew the doors and windows in, continued through the building and blew out the front doors. Another had swept across the golf course, over a filled drainage ditch along O'Leary Avenue, and across the parking lot in front of the base exchange. There it washed away cars and scattered them like toys. The flood then caved in the doors to the base commissary.[9]

Other floods in the housing area knocked in doors and spread the muck inside. Only a few homes were damaged by the sheets of rampant water and ash, and those suffered considerable damage to the possessions inside. However, the roofs of base housing units had held up under the pressure of the ash, rain, and earthquakes. None were collapsed because their spans were

Pyrocastic flows in the Sacobia River came within 400 yards of base housing.

narrow enough to accommodate the extra weight. The barns along the parade field, with their steeply pitched corrugated steel roofs, simply shrugged off the wet ash while their wooden frames flexed with the earthquakes.

Security forces gradually re-established the patrol sectors they had established under their Air Base Ground Defense plan. Eagle directed that no one was to go on the west side of the runways unless they were in a four-wheel-drive vehicle, so the police patrolled open areas and checked houses as they could. It appeared the legion of thieves, who had continually plagued the base, had run like nearly everyone else. Some Philippine Air Force police had never left the base for reasons that are unclear. Possibly, they were never directed out of harm's way, so they stayed. None was injured or lost.

Some thought it looked as though a nuclear war had been fought there. Others thought it looked like a terrible blizzard had swept across the province, while others thought it looked like the moon

Security patrols did not bother to drive on streets; there were not any streets visible under the blanket of ash. They moved through the paths of least resistance. They did not find any evidence of looting either in the housing areas, munitions areas, armory, or the crushed supply warehouses.

In Mactan housing, the area many called Hill housing, the ash was particularly deep. Floods had stacked it two feet high in many places, and patrollers found pumice rocks the size of footballs near the westernmost wall of the base.

Eagle and his command element prioritized their immediate concerns after re-establishing security: water, shelter, food, and communications.

Weight of heavy, wet ash collapsed 110 buildings at Clark, among them this electricity generating plant.

Water and Electricity

One of the initial situation reports (SITREPs) from the base indicated that 600,000 gallons of potable water were available.[10] The estimate was based on water stores that were trapped in large water tanks atop hills on the base. However, the base civil engineers, lead by their deputy commander, Lt Col Sam Kinman,* learned some harsh lessons over the next few days.

Clark's fresh water supply was provided by a series of wells on the base. The wells were pumped by electric motors, and the water moved 24 hours a day into large storage tanks on the hills. A network of supply lines ran from the tanks' gravity-fed water to the homes and work areas across the base. The engineers, in accordance with the plan to evacuate to the college, had trapped water in the storage tanks because they knew electrical power could fail thereby disabling the motors that pumped the wells. They had done their homework well and learned lessons from other bases that had experienced light ash fall from other volcanoes.

What they did not realize, however, was that the network of supply lines that ran from the tanks to faucets and toilets across the base had been damaged by the continuous earthquakes. So, when they released the water from the tanks it quickly seeped out of the damaged sections and was lost. Without electrical power to drive the pumps, there was no way to refill the tanks. Initial estimates were that 15 to 20 percent of electrical power could be restored in 24 hours.[11] The estimate was overly optimistic.

Electricity was not just around the corner. Most of the power-generating plant was buried under wet ash when the roof caved in, and falling branches had severed power lines across the base. The power lines that were still standing were useless until they were cleaned off. The engineers knew the ash was conductive, and piles of it layered the tops of every transformer on electrical poles across the base. Substations and distribution centers were clogged with the stuff. They had only stiff brushes and rags with

* The commander, Col Jim Goodman, was acting as AF Combat Support Group commander at Subic Bay.

which to clean the power distribution system. The enormity of the job dictated the cleaning be done with high-pressure water hoses and large volumes of water.[12]

Joseph Heller would have loved it – the perfect "Catch 22." Without electrical power they could not get water, but they could not get electricity until the water was back on.

Goodman, Kinman, and their men had anticipated the loss of electricity and gathered a supply of emergency generators that could power the wells. However, the wells were very difficult to find because of the blanket of ash across the base. Even with surveyor's maps to guide them, the totally gray base greatly exacerbated the problem. There were no street corners from which to measure. Brass survey markers were buried.

Most eerie of all... was the silence

The civil engineers attacked the problems with a vengeance. SSgt Miguel A. Ley and other power specialists at the power plant worked to restore at least one of the generators to working order. Their biggest problem after cleaning the generator was how to cool it since the big diesel motor that spun the generator was water-cooled by large radiators which laid flat. Once the roof caved in, the ash piled onto the radiators. One of Senior Master Sergeant Burke's workers had covered one radiator with plastic, but the others were open to the elements. It took 50 man-hours to clean the radiator that had been covered, and 150 man-hours to clean the ones not covered.[13]

Even after the radiator was cleaned, though, there was no water pressure to run cooling water through it. So, the power plant people devised an ingenious solution. They brought a water-pumper truck from the fire station to the power plant. There, they manufactured a fitting to attach the fire truck's water supply to the radiator. When they turned the truck on, it cycled water through the diesel's radiator so that the huge generator could produce electricity.[14]

Other engineers started the painstakingly slow process of locating and digging out wells. Ley and other workers hooked up portable generators to them. However, none of the wells had receptacles to which the generators could be easily attached. Ley and his workers had to improvise considerably to hook up the power.[15]

Some started the equally difficult process of locating shut-off valves in order to isolate damaged sections of the water system. It was an extremely time consuming process. The CAT commander assisted by prioritizing specific buildings for water. This "short list" of which facilities needed water helped engineers to determine which valves to close and which to leave open.[16]

Food and Shelter

Food was not a problem. The supply of MREs was virtually unlimited, and although some troops may not have preferred tuna casserole in a brown plastic pouch, no one went hungry. Vast quantities of food were available in the base commissary and the dining areas. Although the refrigerated and frozen foods were quickly lost because of no electricity, a nearly limitless supply of canned goods was available.[17]

Shelter was a more difficult problem. The tent city no longer existed. It was flattened and buried. The base was still being rocked by frequent earthquakes, and every roof held at least six inches of wet compacted ash. Eagle immediately put people to work shoveling the Dau command post's broad, flat roof. It was very slow going. Even though it was only a few hours old, the ash had settled into a hard-pack, and it was back-breaking labor. The roof area of the large Dau building was 10,000 square feet and the ash was compacted to four inches thick. Therefore, the roof held nearly 400 cubic yards of the gritty, wet "sand." The

After volcano eruptions, earthquakes and typhoon, American and Philippine airmen picked their way through the wreckage of the devastated Clark Air Base to raise the US flag once again.

stuff within a few feet of the edge could be shoveled over the side, but the rest had to be carried to the edge before it could be dumped.

Eagle made two rules: nobody sleeps inside a building until the roof is clear, and everybody not on duty guarding the base gets a shovel and goes to work. Enlisted airmen and colonels stood shoulder-to-shoulder on the Dau command post roof moving tons of wet ash slowly to the edge where it could be thrown over the side.

Communications

Communication among the Ash Warriors was superb, thanks to the brick network maintained by the 1961st Communications Group and some innovative phone connections they made. Controls and the central brick station were located in the Dau command post, which had its own electrical generators, so the net never went off the air. The CAT, the wing commander, and Thirteenth Air Force commander were always in touch with the troops. When the decision was made to move to the alternate command post, telephone technicians installed a visual display unit (VDU) and a remote switchboard at Dau. The equipment allowed technicians to control telephone connections and routings on the entire base from a remote location.[18]

Communication beyond the base was a different story. The ash and mud flows disabled the satellite facility that was the nerve center of military communications to the outside world. Fortunately, the communications commander foresaw the possibility of losing use of the satellite facility* and installed an alternate piece of equipment in the Dau command post.† The alternative system could not handle the same volume of messages as the primary, but throughout the days after the major eruptions, the base was still able to receive and transmit official message traffic.[19]

Security

By nightfall on June 16, the small security force that returned to Clark was bedded down in the Dau command post. They went to bed very tired and very dirty. Water buffaloes,‡ some of which had been filled from the fire department's cache of chlorinated water, were stationed at strategic locations. The water was only for drinking, however, and bathing with potable water was not permitted. The gritty ash coated their bodies under their BDUs. The abrasive stuff rubbed ankles raw inside their boots and reddened skin under their belts. It caked in their hair (some called it "cement-head"), and many wrapped their heads in bandanas, motorcycle gang-style, to keep the stuff out. The abrasive ash was particularly hard on the security forces' weapons. There was no way to handle the weapons without rubbing the ash on the metal finish. The abrasion quickly wore away protective lubrication and the weapons started to rust in the humid tropical air.[20]

* A not uncommon occurrence during typhoons.

† TEQCOM Z-200.

‡ Water carriers mounted on a small, two-wheel trailer.

Shovels and brooms are used to clear the roofs of remaining buildings of tons of ash.

By the end of the day on June 19, the entire mission essential force had returned to Clark from the college.[21] Eagle brought them back in manageable increments. Security police and engineers who could work the electrical and water problems returned first. Support personnel followed. Everyone pitched in to shovel roofs. Clever carpenters constructed make-shift "ash pushers" from plywood and scrap lumber to make the roof clearing operation faster and easier. The CAT submitted an emergency request for snow shovels. One officer commented that he shoveled ash until he "could not stand up straight."[22] Their efforts were more than busy work; buildings and roofs continued to collapse from the stress of the heavy ash and the shaking from earthquakes that frequently rolled through the base.

The mounted horse patrol (MHP) troopers returned to find that all their horses, as well as the recreational horses, had survived. The stable buildings, which had relatively small roofs, had not collapsed. The horses were not in good condition, though. Some Filipino stable hands had returned to feed and water the horses, but there were not enough hands to get each of the nearly 100 horses out for exercise, so the terrified animals were stiff and lame from standing in their stalls for nearly five days.[23]

MHP riders acted immediately to rejuvenate the horses. Even though there was little water on the rest of the base, it flowed freely at the stables. The stables were at a low area near the end of the water system. Water was trapped there, and the MHP used it to water and wash the horses as well as themselves. Fodder was a different matter. The horses' grain feed had to be supplemented by grazing the generous pastures around the stables, and the pastures no longer existed. SSgt Tony Price and others started an all-out effort to find enough hay to feed all the horses. For several days, the MHP spent every daylight hour walking the horses up and down a path where the road had once been to get the horses ready for riding.[24] Also, they "recruited" several horses from the recreational stables next door to become MHP mounts.

The frequent black ash clouds that drifted through made it look like the dark side of the moon

Clark security forces also had over 100 military working dogs assigned to their unit. When the order was given to retreat to the colleges on June 15, only half the dogs, the ones on duty, were taken to the college. The other half were left in their kennels with food and water. One dog that was left behind did not survive. It was buried in a large police dog cemetery that held the remains of other dogs that had served well over the decades.[25]

Across the base, security policeman continued to "eat ash" 24 hours a day. The Air Base Ground Defense scheme continued to work well, and thefts on the base dropped to a third of the level that existed prior to the evacuation of dependents and non-essential military personnel on June 10.[26] The security police commander, Gladiator, attributed the success of his police to four factors. First, identifying intruders was easier. Prior to the evacuation, there were thousands of residents, domestics, and other workers on the base. After the evacuation, if a patroller saw someone who was not wearing BDUs walking in the housing area or near a supply warehouse, that person was not supposed to be there. Second, dog handlers were given permission to unleash their dogs to pursue intruders. Third, the entire 3d Security Police Group was present for duty, and they all worked 12-hour shifts. There were no distractions such as training meetings, dental appointments, or family matters to reduce the on-duty force. Fourth, the ABGD scheme put the same patrollers in the same area every day or night. The police learned their special turf well, knew likely intrusion routes and could see when something was out of place.[27]

One security sector, in the Hill housing area closest to the volcano, nicknamed itself Dark Side Control. The nearly continuous ash fall in the area and the frequent black ash clouds that drifted through made it look like the dark side of the moon. The unit set up operations in a base house that had been unoccupied before the evacuation. The Dark Side Control commander, Capt Richard Scott referred to his 58-man contingent as "very motivated and highly aggressive patrolmen."

Dark Side Control, as well as many other control points, maintained a menagerie of abandoned pets. Many evacuees, who went to Subic Bay on June 10, left pets in their quarters, assuming a quick return to the base. Although residents who left pets behind, put out food and water, those supplies quickly ran out. Police patrolling the 3,000 houses, released many of the pets they discovered to fend for themselves, but some were adopted by the Ash Warriors. Dark Side Control had a black Labrador retriever, a couple of rabbits and a parrot.[28] Unfortunately some pets died in their homes of starvation or dehydration.

Hot Meals

During the week after the eruption, a small contingent of services personnel worked to

Heavy erosion from rushing water was common throughout the base.

restore dining facilities. The Tropical Inn, which was the main enlisted dining hall, the officers' club, and the non-commissioned officers' club all sustained significant damage from wet ash. A new airmen's club, which was opened just before the evacuation, was not as badly damaged, but needed extensive cleaning.

Workers opened a small eating facility, the Express Diner, near the flightline. TSgt Dorothy Lewis and six other providers of the Air Force Commissary Service shuttled canned goods from the commissary warehouse to SMSgt Donald Ostrander and his crews at the Express Diner. The two units, working together, took pride in the fact that they could go from "cans opened to [hot] meals served in an hour and a half."[29]

"Most people I talked to don't ever want to remember what it felt like"

Vehicles crowded into the parking lot, and a long line waited outside the Express Diner for breakfast, lunch, and dinner. The parking lot was piled high with ash along its sides where earthmovers had piled it like dirty, gray snow that would not melt. Troops waiting for a hot meal shuffled along the line. When they came inside, they stomped the ash from their feet and went through the cafeteria line as food servers piled paper plates with main courses, hot vegetables, and mashed potatoes – a much welcome respite from MREs.

Crow Valley, former site of the Cope Thunder exercises (top), was filled with pyroclastic flow (bottom).

A bright spot in the gray ash was MSgt Judy Sanders, who was ordered back from Subic Bay along with others who had special skills. Sanders was a specialist at operating club facilities such as officers' clubs and enlisted clubs. The club facility least damaged was the new airmen's club, and Sanders immediately went to work hiring Filipino laborers to clean out the place. It was no easy task. Muddy ash soaked the carpeting, and the walls started to mildew in the humid atmosphere. Nonetheless, Sanders and her worker bees opened the airmen's club in three days.

Ash Warriors stomped ash off their feet, brushed it as best they could from filthy uniforms and walked into a clean, dry place. There they could enjoy a cold San Miguel, have snacks and listen to Jimmy Buffett sing on the juke box, "Where ya gonna go when the volcano blows." Electricity came from an emergency generator behind the building. Three days later Sanders opened the kitchen at the airmen's club, and the mission essential force had, within the club, a touch of the way things used to be at Clark.[30]

Getting Clean

The civil engineering squadron continued to struggle with electrical power and water. For two steps forward there seemed to be one step backward. Eagle was deeply concerned about getting the water on line for two reasons. First, none of the toilets could be flushed, so troops were using the few portable toilets that could be found as well as slit trenches for latrines. Second, everyone was filthy, which was not only bad medically, but the never-ending dirt had a deleterious effect on morale.

Grime said, "I was trying to get . . . water up so they could take a shower, you can imagine ash, how it cakes on you after a time, and being able to use a toilet. It's something . . . people don't think about, but concerns with sanitation could shut down that whole operation faster than anything else. We had the darndest time, of all things . . . keeping people out of latrines. They wanted some privacy. We were quickly as possible cleaning up port-a-potties, but frequently a lot of people used latrines that weren't working, and that causes a mess in terms of hygiene. And, I remember one night sitting in [the Express Diner]. We were able to get hot food going, but we couldn't get enough power using the emergency generators to get air conditioning, and I was watching the security police come in.

I could look in their eyes and I could see they almost had it. They had that stare like we ain't got many days left and then we're going to start giving up . . . and that was [one of] the things that scared me the most."[31]

Another Ash Warrior recalled, "Most people I talked to don't ever want to remember what it felt like. You couldn't get clean. Eating MREs wasn't a chore . . . shoveling ash wasn't a chore. But, no flush toilets, no running water to take a shower, no water to brush your teeth with, no water to rinse your glasses with, for 10 or 11 days, that got to be old after a while."[32]

Some troops were innovative. About a week after returning from the college, Lucky was doing busy work in the alternate command post just before dawn when he saw an airman walk by who had clean hair. The CAT commander jumped up from his desk, caught up with the clean-hair man in the hall and asked him if he had taken a shower someplace. The airman said yes and pointed the colonel in the right direction. Lucky grabbed a bar of soap, clean clothes and his only clean towel from his bugout bag and headed out of the building. Across an ash-covered rice paddy a quarter of a mile from the command post, the colonel found a small farmer's hut. The hut was abandoned, but behind it was a pump sticking up out of the ash, and a bucket half full of water. Written on the side of the plastic bucket, in grease pencil, were the words, "Enjoy, but leave water for priming." He carefully poured the water in the top and started pumping the handle up and down. In a moment clean water started pouring out of the pump. He stripped naked and poured the first bucket of water over his head and then another and another and another. Each one felt better than the first. He stood there for 10 minutes, naked in the middle of a rice paddy just pouring clean water over his head. Finally, he used the soap to scrub thoroughly, dried off, dressed, and left a bucket of water for the next person. By mid-day there was a long line at the rice paddy pump.[33]

Cold Beer

Prior to the June 10 evacuation, the area of Angeles City just outside Clark's base was bustling with shops, sari-sari stores, restaurants, and bars that were frequented by the American service men and women and their families. A large liberty area was designated in the city, and the city outside the liberty area was off-limits to all Americans. It was off-limits because General Studer did not believe US and Philippine security forces could protect Americans from the terrorist threat unless the area they had to protect was of a manageable size. The liberty area, when viewed on a map, was in the shape of a fish. A few days after the Ash Warriors returned from the college, General Studer approved liberty in the fish for everyone.

The word quickly spread outside the gates, and when the troops went out the gates, a few enterprising night club and bar owners had hooked up generators to their lights and beer coolers to accommodate their patrons. However, the fish was even more devastated than the base. Ash clogged the dark streets and many buildings were damaged badly. The vitality of the city was gone, and for many, drinking a cold San Miguel at their favorite watering holes was bittersweet.

Still Struggling for Running Water

Using an improvised electrical system of emergency generators and fire trucks hooked to main generators, the engineers were able to start pumping water from the wells into the storage tanks. It was a slow process because the pipes in the water supply system held six times as much water as the tanks. So, they had to fill the tanks several times before there was any significant pressure in the lines. Throughout the process the volcano refused to cooperate. Every time it erupted and ash blew across the base, some emergency generators and main power lines shorted out as the conductive ash blew into the systems. Everything had to be cleaned of ash before the process could start again.[34]

Ash-covered Chambers Hall, a landmark familiar to everyone at Clark AB.

Grudgingly, the water system yielded to Goodman's engineers, who had to dig some water control valves out of collapsed buildings. Fortunately, Staff Sergeant Walla, the NCOIC* of the plumbing shop, and others had maintained accurate maps of valve locations.[35] Finally, water pressure sprouted showers in selected buildings.

The largest building was Chambers Hall. Chambers was a six-story, 300-plus room dormitory in the center of the base. New officer arrivals stayed at Chambers until housing was available. Officers on temporary duty at Clark stayed there. Countless thousands of officers, who lived at or traveled through Clark, remember it as a home-away-from-home. Services people hired Filipino workers to shovel the flat roof of the hotel clear of ash, and by the time they were finished, electricity and water were available, so the building was opened to occupancy. Officers who had been sleeping in their cars at the Dau command post, gathered their few belongings and checked into Chambers for clean sheets and a real shower.

The eruption of Mount Pinatubo was the world's largest in more than half a century and probably the second largest of the century

Goodman's engineers had checked the building thoroughly for damage, and there was none. The building was constructed to withstand strong earthquakes, and its expansion joints had merely flexed as the temblors rattled Clark. The motion of the building, especially on the sixth floor, sometimes threw occupants from their beds, so the top floors were abandoned. Less than 10 days earlier, people who slept in their cars rather than in the two-story buildings at the college, blithely slept well under several floors of intermittently swaying concrete.

Aerial view of Base Supply showing the incredible damage sustained by the combined efforts of both volcano and typhoon.

Ash, Ash, Everywhere

During the few times that airborne ash cleared out with prevailing winds, helicopter operations resumed carrying the volcanologists and General Studer to view the crater and what was happening inside it. The flights were sometimes tricky. Using the call sign "Cactus," the flights were conducted almost exclusively during daylight hours when ash clouds could be seen and avoided. The base's three UH-1 helicopters were hangared in the former alert facility at the south end of the flight line to protect them from unexpected ash fall. Large pads were cleared to provide as clean an area as possible for the aircraft to takeoff, hover, and land. In addition to a pad directly adjacent to the hangars, another large pad near base operations was also cleared to allow helicopters from off-station to operate into and out of the base without interfering with Clark's birds. The aircraft were rinsed daily by maintenance personnel to help prevent corrosion.

Even though the large pads were cleared of ash, they still had to be wetted down with fire hoses to prevent ash from blowing during takeoffs and landings. The pilots used maximum performance takeoff procedures whereby they made a vertical takeoff to 50 to 75 feet before they transitioned to forward flight. Landings were made from a steep descent to reduce the time the aircraft might encounter blowing ash from the rotor down-wash. Wind indicating devices were placed near each pad to insure wind direction was discernible to aircrews. The pad that was cleared on the parking ramp near base operations presented more blowing ash problems because of vehicular traffic around the pad. As the ash dried out, vehicles operating near the pad created a great deal of dust which blew toward the pad. The pad located near the hangar had no such problems because it was far from vehicular operations.[36]

The volcano continued to erupt throughout June, sending huge columns of ash into the air. The Ash Warriors, callused from the experience of "historic" eruptions, barely took time to watch. However, the CAT continued to watch the scientists watch the mountain and made plans to bug out again, if necessary. All cars were to be parked facing out to avoid snarls in parking lots clogged with ash and abandoned

* Non-commissioned officer in charge.

vehicles. No one was to sleep in an area unless one person was awake with an operable radio. Guards were posted in dormitory sleeping areas to alert occupants. If the siren blew everyone was to proceed by vehicle to the main gate and await further instructions.[37]

The volcanologists worked to define the scale of the eruptions, an effort that continued for several years and culminated in the publication of *Fire and Mud*, edited by USGS-PHIVOLCS volcanologists Christopher G. Newhall and Raymundo S. Punongbayan.

The eruption of Mount Pinatubo was the world's largest in more than half a century and probably the second largest of the century. It ejected more than five cubic kilometers of magma, an order of magnitude larger than the Mount Saint Helens eruption in 1980. Only the 1912 eruption of Novarupta in Alaska was bigger than Pinatubo within the twentieth century[38]. The ash, combined with rainfall from Yunya, blanketed 7,500 square kilometers of Luzon at least a centimeter thick. Almost the entire island received some amount of ash fall.[39]

Over 100 buildings collapsed at Clark, and another 500 were damaged, most of them extensively. Within the Filipino communities that surrounded the American bases, between 200 and 300 died, most of them from collapsing roofs. Without Tropical Storm Yunya, the death toll would no doubt have been less.[40]

Voluminous pyroclastic flows extended as far as 7 to 10 miles from the summit down all watersheds and impacted an area of 240 square miles, "dramatically modifying the pre-existing topography." Some filled existing valleys to a depth of over 600 feet.[41]

The small lahars that swept across Clark the afternoon of Black Saturday caused little damage relative to what happened in the river beds north and south of the base. The most significant damage happened in the Abacan River south of Clark. Dwellings upstream from Sapang Bato* were inundated. Many buildings and all bridges in Angeles City were destroyed that day, or fell to lahars over the next few months. The Northern Expressway bridge, on the main road to Manila, also fell as lahars thundered down the river like shiploads of wet cement spiced with boulders.[42]

Nonetheless the moment was not without its bright spots. Personnel Bobbitt, who always seemed to be talking on the radio as he scrounged supplies, was pleasantly surprised on June 23. At 10:15 a.m., General Studer and Colonel Grime STEP† promoted him on the spot to staff sergeant. He was promoted for "his sustained exemplary performance in the service of the Clark family." The commanders specifically noted Sergeant Bobbitt's efforts during the previous two weeks.

Bobbitt, uncharacteristically speechless, said afterwards, "I'm in a position to help, and I'll continue to do that until we run out of supplies."[43]

What About the Families?

One of the main concerns of every Ash Warrior was the welfare of his or her family. Specialists who returned to Clark from Subic Bay told grim stories about the conditions at the Navy bases. The ash fall at Subic Bay was at least as bad as at Clark, and many buildings on the Navy bases collapsed as well. There were critical shortages of food and water. Clark's newspaper, the *Philippine Flyer*, assured the mission essential team that the families would be cared for and evacuated to the United States, but the Ash Warriors could not know how bad it really was for their families until they were reunited and heard their incredible stories first-hand.

** A village just outside a gate on the southwest side of Clark.*

† Stripes to Exceptional Performers program. An AF program that gives local commanders the authority to promote a few airmen each year outside the normal board selection process.

NOTES TO CHAPTER 6

1 Interview, William Dassler, Col (Ret), former commander of 3d Security Police Group, December 7, 1998 [hereafter cited as Dassler interview].

2 Ibid.

3 JTF-FIERY VIGIL SITREP, 161915Z JUN 91.

4 Grime interview 1992.

5 JTF-FIERY VIGIL SITREP, 161915Z JUN 91.

6 JTF-FIERY VIGIL SITREP, 161015Z JUN 91.

7 Fire and Mud, p 76.

8 JTF-FIERY VIGIL SITREP, 161915Z JUN 91.

9 Ibid.

10 JTF-FIERY VIGIL SITREP, 171451Z JUN 91.

11 Ibid.

12 JULLS Long Report, JULLS No. 73157-04215.

13 JULLS Long Report, JULLS No. 73159-39818.

14 JULLS Long Report, JULLS No. 73150-84568.

15 Philippine Flyer, Special Fiery Vigil Bulletin No. 8, June 24, 1991, p 1.

16 JULLS Long Report, JULLS No: 73150-18877, August 15, 1991.

17 JTF-FIERY VIGIL SITREP, 171451Z JUN 91.

18 JULLS Long Report, JULLS No: 73057-72998, August 15, 1991.

19 JULLS Long Report, JULLS No: 72954-5986, August 15, 1991.

20 Dassler interview.

21 Philippine Flyer, Special Fiery Vigil Bulletin No. 4, June 20, 1991, p 1.

22 Rand interview.

23 Mikkelson interview.

24 Ibid.

25 JULLS Long Report, JULLS No. 73134-55165.

26 JULLS Long Report, JULLS No. 73133-94911, August 15, 1991.

27 Dassler interview.

28 Philippine Flyer, Special Fiery Vigil Bulletin No. 9, June 25 1991, p 1.

29 Philippine Flyer, Special Fiery Vigil Bulletin No. 11, June 27, 1991, p 1.

30 Philippine Flyer, Special Fiery Vigil Bulletin No. 6, June 22, 1991.

31 Grime interview 1992.

32 Rand interview.

33 Author's journal.

34 JULLS Long Report, JULLS No: 73150-18877.

35 JULLS Long Report, JULLS No: 73161-14294, August 15, 1991.

36 JULLS Long Report, JULLS No: 80545-04093.

37 Philippine Flyer, Special Fiery Vigil Bulletin No. 7, June 23, 1991, p 1.

38 Fire and Mud, p 18.

39 Ibid., p 14.

40 Ibid., p 15.

41 Ibid., p 15.

42 Ibid. ,p 16.

43 Philippine Flyer, Special Fiery Vigil Bulletin No. 7, June 23, 1991, p 2.

Chapter 7 THE EVACUEES

During Black Saturday, June 15, 1991, American service men and women and their families huddled in shaking, creaking buildings at Subic Bay, Cubi Point, and San Miguel. For some, their worst nightmares had come true. First, the evacuation order from Clark ripped them from their homes when many were unprepared. Then they were subjected to long lines in the hot sun at Clark and Subic Bay, not to mention hours of bumper-to-bumper, stop-and-go traffic on the highway to Subic Bay. Conditions at the mobbed Navy bases were nothing short of chaotic. Vehicles were stuffed by the hundreds into huge fields. Lines for housing snaked out of buildings into the scorching tropical sun.

However, in the spirit of what many describe as "the military family taking care of each other," the evacuees and their reluctant hosts attempted to achieve a measure of normalcy on the overstuffed Navy bases. The Clark Air Base Wagner High School Class of '91 graduated on June 13 in a makeshift ceremony in the Subic Bay auditorium, while an announcement told other classes of the cancellation of final examinations. Although the circumstances of the graduation were not what the graduates may have imagined, the ceremony is fondly remembered by those evacuees who attended as a touch of home that cheered many.[1]

The forces of nature overcame the best human intentions, though, and the huddled masses cowered as tons of wet ash fell, earthquakes smashed through the bases and buildings collapsed. No matter how happy a face they tried to paint on the situation, everyone knew they were at the mercy of the elements with absolutely no control of the outcome. Many took to their beds to seek respite in sleep. One teenager recalled, "If it was going to kill me, it was going to kill me. At least I would die in my sleep under my own covers."[2]

Joint Task Force-Fiery Vigil

Although it seemed as if they had been abandoned to nature's brutal realities, the military machine was already working to relieve the suffering. On June 10, the day Clark evacuated and five days before the cataclysmic eruptions, the United States Commander in Chief, Pacific (USCINCPAC), a Navy four-star admiral in Hawaii, activated a joint task force to protect American lives and property in response to the volcano.[3] The task force was nicknamed Fiery Vigil. If the title was intended to be humorous, there is no record of anyone at Clark or Subic Bay finding amusement in the appellation.

Formations of such task forces, called "joint" because they involve all military services, is standard procedure during a crisis. It allows the senior commander, in this case USCINCPAC, to establish a chain of command with which to attack the problem directly. It also brings to bear the resources of all the military services in a coordinated effort. As luck would have it, the Joint Chiefs of Staff had executed a no-notice interoperability exercise with General Studer as the commander of the Joint Task Force-Philippines on April 30, 1990. General Studer's task was to execute command, control, and communications (C3) responsibilities over a joint air, ground, and naval force. The exercise identified shortfalls in C3 and the lack of a consolidated non-combatant evacuation operation (NEO) plan that could support all on-island evacuation efforts. As a result, the state-of-the-art brick system was installed at Clark, and, fortu-

Goodbye Dad – Dustin Van Orme waves goodbye to his father after boarding a bus to join other evacuees who are boarding ships to leave Subic Bay.

itously, a comprehensive NEO plan was developed.[4] When it came time to pick a commander for Joint Task Force Fiery Vigil (JTF-FV), USCINCPAC passed over the Navy admiral in the Philippines and heeded the suggestion of US Ambassador to the Philippines, Nicholas Platt, to appoint Studer as commander. Apparently Platt thought Studer had done well during the recent exercise. USCINCPAC also dispatched a staff of officers from Hawaii to assist Studer. Many were the same officers who had served during the exercise. As commander, all military in the Philippines were subordinate to Studer including all Navy personnel and families at Subic Bay.[5]

General Studer and the senior Navy officer at Subic Bay were in agreement by June 12* that Subic Bay was so overloaded that an evacuation back to the United States was the only possible course of action unless the volcano rapidly reversed its course. Since Pinatubo was in a constant state of eruption, there was no possibility of returning the evacuees to Clark, and Subic's facilities, especially the water supply, were stressed to breaking.[6] General Studer determined that he would keep 1,500 people at Clark (1,000 security police plus 500 support and maintenance) and leave a large manpower pool of 5,000 at Subic Bay. Therefore, as the danger from the volcano subsided, selected groups of workers who had critical skills could return to Clark as needed. However, Studer was laying the groundwork for a decision he thought inevitable. It was clear to him that not only would the dependents need to be evacuated, but all of the Clark military huddled at Subic Bay would have to go as well.[7] In a situation report to USCINCPAC on June 12, General Studer said, "I do not believe we can return to any semblance of normalcy in the near future and leaving 11,500 Clark people at Subic for more than a few days, certainly not weeks, is out of the question."[8]

Weary travelers evacuating through Andersen AB, Guam.

After the above message was sent, the volcano erupted twice in 24 hours, and General Studer sent the following words in a message the next day to USCINCPAC and officials in Washington, DC, "The large number of evacuees has stressed the resources of the Subic Bay complex to near maximum. The capability to provide a minimal living standard for the evacuees for a sustained period is not considered feasible. The further eruptions of Mount Pinatubo [exacerbate] the situation by denying the possibility of returning to normal operations at Clark Air Base for up to six months. As a consequence, my first priority is to get all Air Force dependents out of the Philippines. All planned near-term actions are directed toward that goal."[9]

The JTF-FV staff, lead by Studer's deputy, Col Bruce Freeman, devised an orderly flow of non-essential military service members and all families out of the Philippines. They established three categories.

Category 1: Anyone with a June port call† whose household goods were already packed and shipped. This category was to leave as soon as possible. The staff was coordinating for commercial wide-body aircraft to airlift these people from either Manila or Cubi Point.

Category 2: Those with port calls after 1 July would have their tours shortened.

Category 3: All dependents. Studer requested permission to initiate the early return of dependents. He said, "Because of heavy support infrastructure this group requires, request authority to commence within 10 days."[10]

The USCINCPAC crisis action team transmitted a message to the Secretary of Defense requesting permission to declare an emergency evacuation of the Philippines. General Studer agreed and asked that the evacuation of dependents and non-mission essential personnel start prior to June 17. Two planeloads of people in Category 1 flew from Cubi Point to Andersen Air Force Base, Guam, under this scheme and before Black Saturday. A regular schedule of flights from Manila and Cubi Point was approved.

* Three days before Black Saturday.

† A port call is a scheduled departure.

Further, US Transportation Command approved one commercial aircraft per day to transport 200 to 250 passengers from Cubi Point to Travis Air Force Base beginning June 14. Each passenger was authorized three to four pieces of baggage. Ten to seventeen pets were also authorized on commercial flights depending on size of containers. Five missions were requested over and above the normally scheduled twice-a-week chartered missions.[12] However, the volcanic eruptions of June 13 and 14 overcame JTF-Fiery Vigil's best intentions. Cubi Point airfield was closed most of the time because aircraft could not fly through, or even around, the large clouds of ash in the air.

On June 15, as Pinatubo convulsed in the paroxysms that would turn day into night, a regularly scheduled Boeing 747, loaded with evacuees, departed Cubi Point for Kadena Air Base, Japan. During its climb out, ash from Pinatubo snuffed out one of the jumbo-jet's engines. The skillful crew regained use of the engine and landed uneventfully at Kadena, but that was the last airlift of evacuees from Subic Bay.[13]

It was clear that not only would the dependents need to be evacuated, but all of the Clark military at Subic Bay would have to go as well

When it became obvious the Cubi Point runway was no longer a viable option for use during the evacuation, other options were considered. One option was to move them to Manila by ship then out of Manila via aircraft. However, Manila was so close to Pinatubo that ash would likely affect Manila flights as well. Planners looked around for another airfield to which ships could take the evacuees to meet up with air transportation. Maj Jim Simpson, the 834th Pacific Airlift Control Element* Chief recalled a former USAF base, used during the Vietnam war, at Mactan on the island of Cebu, which was about 300 miles south of Subic Bay and Luzon. A quick check of air navigation charts in the area showed an operational airstrip there. The 834th CAT made this information available to the PACAF and PACOM CATs. USCINCPAC chose the old airstrip at Mactan since Aquino Airport at Manila would probably be closed due to ash fall for several days. The PACOM CAT estimated the evacuation would total 20,000 with the first ship departing Subic Bay on 16 June.

Evacuation Armada

The aircraft carrier USS *Abraham Lincoln*, on her maiden voyage, and her battle group, along with ships from two amphibious ready groups, diverted from their course to the Indian Ocean to Subic Bay. The carrier *Midway* and guided missile cruiser *Mobile Bay* were ordered from Yokosuka, Japan.[15] Nearly 48 hours before Black Saturday happened, Navy ships were on the way to Subic Bay. The Navy was ahead of the game, and its quick reaction saved many lives.

The final evacuation plan, approved by USCINCPAC, was for Navy ships to enter Subic Bay and load with as many evacuees as they could safely carry. The ships would then sail to Cebu (a trip of 24 to 36 hours) where they would leave the evacuees and return to Subic for more evacuees. The evacuees would be put on aircraft at Mactan, the old Vietnam-era airfield, and then flown to Andersen Air Force Base, Guam. At Guam the evacuees would be processed for return to the US mainland while they were given some time to rest and clean up. Then the evacuees would be flown to Travis Air Force Base, California, Norton Air Force Base, California, or McChord Air Force Base, Washington, where they would be given commercial airline transportation to a "safe haven"† location of their choice within the contiguous 48 states.

It seemed a solid plan, but as in any effort, the devil is in the details. As far as the Clark evacuees were concerned, the ships could not get there fast enough. Their ranks were about to swell with several thousand Navy service members, spouses, children and pets. The support infrastructure of Subic Bay was as badly damaged as Clark's, since there was no electricity and, therefore, limited fresh water and food.

The first people to be evacuated were the ones who had port calls in the month of June; these evacuees had already been processed and were ready to leave on short notice. The USS *Arkansas*, and frigates *Rodney M. Davis* and *Kurtz* left Subic on June 16 with 860 such evacuees.

On the same day, June 16, the Secretary of Defense formally approved General Studer's request to remove all non-essential military personnel and dependents from the Philippines under emergency evacuation orders.

* *At Hickam Air Force Base, Hawaii.*

† *An official designation under emergency evacuation orders, i.e. a place to live temporarily until a permanent assignment is tendered.*

The Interminable Lines

Evacuees lined up in front of the Sampaguita Club to get a berth on the ships. At first the biggest problem was getting the word out that an all-out evacuation of the naval facilities was underway. However, the network the first sergeants had established carried the day and a slow trickle of evacuees looking for a way out of Subic Bay became a stream and then a flood as the word spread. Eventually, the Subic Bay FEN radio station resumed broadcasting the departure schedules.

Conditions at Subic were desperate and becoming worse by the hour. Food was difficult to find. Drinking water was impossible. Some Navy services people made well-intentioned efforts to distribute MREs, but their efforts were ineffective with thousands of people looking for food and water.

A US Navy dependent wife, who had opened her home to 30 Clark evacuees, recalled the conditions, "It looked like a gray snowstorm had hit. Power lines were down everywhere. On Monday [June 17] there was so much traffic, and no electricity. On Tuesday, there was no food, gas, nothing. I was drinking water out of the waterbed. Outside people were fighting for food, and siphoning gas."[16]

Col Jim Goodman, Clark's Base Civil Engineer, was acting as Combat Support Group Commander at Subic. He spent most waking hours walking the lines along the docks. His biggest concern was controlling the panic that was slowly working its way to the surface as the lines got longer and the food and water more scarce.[17]

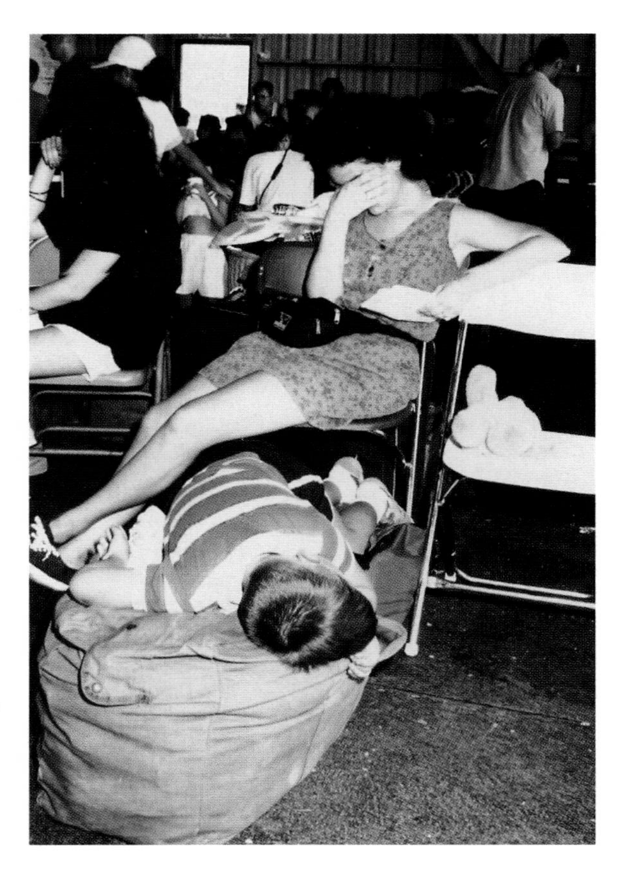

Exhausted, a young mother and her son try to rest in a hangar at Andersen AFB, Guam, during the evacuation.

The *Lincoln* left Subic on June 17 with more than 2,000 evacuees. It was the last of the battle group to sail that day and was preceded by the *Lake Champlain* (300 evacuees), *Long Beach* (572), *Merrill* (298), *Gary* (309), *Inghram* (282), *Roanoke* (500) and *Passumpsic* (192). All told, 4,300 persons sailed for Cebu.[18]

Goodman and others were frustrated by a huge bottleneck in the processing caused by US immigration and customs officials. These officials were required by their regulations, as they interpreted them, to document thoroughly the citizenship of each person boarding a US ship. Of course, many evacuees had not taken the proper identification as outlined in the evacuation pamphlet. So, the back-up in front of the immigration stations seemed interminable to Goodman. It seemed to him that his pleas to speed the process fell on deaf ears, and the lines in the hot sun never got smaller.

Goodman remembered an incident that clearly describes the situation. "After a week of loading folks on ships, long processing lines, longer days and few of the comforts, we were about to send the final 2,000 evacuees to the USS *Midway* for their trip to Cebu, the first leg of a long journey home. The processing was to begin early at 0600 so we could avoid the heat of the day. Additionally, our one water buffalo was nearly dry, and despite my pleas to the Navy command post and Admiral Mercer's staff, no water was forthcoming. The entire base appeared to be dry. To my extreme frustration, the processing line moved very slowly. The immigration people were being a real pain in the ass since some questionable folks had slipped through their grasp in the previous days and they were determined to right the wrongs, exactly when we needed a break the most.

"Nine a.m. arrived and we had made little progress. The sun was bearing down, and the water was gone. Then, a minor crisis occurred that had all the earmarks of a major event. Eleven buses arrived from [San Miguel housing] area. The officer in charge of that location had apparently told the pregnant women and those with children that when they got to Subic Bay, they should immediately proceed to the front of the processing line. Incredibly, as the buses emptied, it looked like the Oklahoma land rush. Every mother with a child under 20 was headed for the front of the half-mile long line. A captain in a flight suit met me where the lemming trails came together and between the two of us, we tried to keep a full-fledged riot from developing. After a few very long minutes, a number of officers and senior NCOs joined us and calm prevailed.

"All of this frustration simply set the stage for what was about to develop. As several other officers, mostly flight surgeons and chaplains, and I walked the line repeatedly, trying to prevent panic from developing, I was getting serious pleas from the docs that dehydration would soon claim some of the young children if we didn't find some water. Several more calls to the Subic engineers revealed that there was no supply of water. Despite the apparent crisis, folks in the line were unbelievably calm, trusting, I'm sure, that we would find a way. I was absolutely out of ideas when I felt a tug at my sleeve. It was one of my sergeants, Jim Stogner. He said, 'Colonel, what can I do?' All I said was, 'Find water, Jim.' I didn't really believe it was in the cards."

"After about a half hour of trying to get the children out of the sun, trying to get the immigration people to loosen up, and trying to offer encouragement to a lot of concerned folks, I once again felt the tug at my sleeve. Jim was there again – I swear he had wings. He said, 'I've got water, it's tested, and we're ready to go.' Seems he had discovered several Filipinos washing down electrical connectors with a 2,500-gallon tank truck. He simply jumped in and drove the truck away. After arriving at our location, he ran a quick chemical test on the water to insure that it was safe and ready for consumption.

"It was easy for me to get about 50 volunteers to distribute the water. Within 10 minutes, everyone in the line was watered down and a real crisis was averted. Shortly thereafter, the processing line started to move more quickly, and by late afternoon, I got to shake a lot of tired hands and see some amazing smiles as those tired folks scrambled up the Midway's gangplank. I often wonder how many lives Jim Stogner saved that day."[19]

Another lifesaver was Air Force Nurse Debbie Head who organized a relief station for infants in a nearby bowling alley. Mothers of infants could leave the line and bring their infants to Head's makeshift clinic. There, the children could get out of the sun and be checked for dehydration.[20]

One teenaged, dependent daughter recalled going to line up to get on the *Lincoln*. "We threw all of our things into our bags, threw the cat into her carrier, and hauled ass for the bus. I didn't think we were going to make it and I almost started to cry. I was pretty sure that we were going to die in that horrible place. We made it to the bus. It was quite an experience . . running through what seemed like a million feet of ash and mud, dragging our bags and the poor traumatized cat. If you've ever run for your life before, then you understand. It was do or die. We loaded onto the bus and it drove away about two minutes later. It was such a relief just to look out the back window and just leave all of the devastation behind. I felt like I could breathe again."[21]

Subic Bay's living areas, streets, and passenger processing areas were littered with personal items the evacuees could not carry. A typical evacuee family consisted of a young mother, often Filipina, with one or two toddlers and a family pet. Few men traveled in the first waves of evacuees because 1,500 Ash Warriors were at Clark, and another 5,000 were identified to remain at Subic. This decision was quickly reversed when it became apparent that Clark could not support such a large number. However, the result was that many hundreds of families left on ships from Subic Bay while their spouses were in the same evacuation stream two or three days behind them.

Even though some families were separating, others were starting their first days together. On June 16, 313 airmen made application to the American Embassy for marriage licenses. Most of these were approved quickly, and on June 18 the number of marriage applications rocketed to approximately 850. Once married, the brides

The stunned and exhausted look on this young evacuee's face tells the story of her ordeal before she arrived at Andersen AFB, Guam.

were allowed on the ships to evacuate to the United States where they would have 60 days to complete the paperwork that would make them citizens. Of course, a few of these marriages were between US citizens, but the vast majority of them were weddings of US servicemen to Filipinas. Some were probably not eligible for immigration, but US officials agreed to approve all the weddings as an acceptable price for order, safety, protection of US dependents, and the rapidly deteriorating conditions.[22]

Evacuees were allowed to take whatever they could carry onto the ships. Families left cars packed with personal items. They were instructed to leave their car keys with their first sergeant or at the Air Force processing center. Many did not get the word or elected to not do so. Every trash container between living areas and the processing station was piled to overflowing with more personal items the young wives could not carry. The wife of one senior officer, who walked the lines giving encouragement, was dismayed to see a wedding photo album thrown in a trash can.[23] A young wife, in helpless tears, tried to manage three small boys, their baggage, and a large dog. She sobbed her frustration to General Studer's wife, whom she neither knew nor recognized, about the uselessness of trying to take the dog. The young woman's husband, who was still at Clark, had only recently acquired the rambunctious animal as a guard dog for their house. Finally, a colonel's wife, who was patrolling the lines and helping evacuees, took the dog from its distraught and reluctant owner and delivered it to the veterinarians' tent, where she was surprised to find that the veterinarians had no apparent plan to deal with animals that were left behind. The dog was last seen tied to a tent post, barking as the mother struggled down the line with her sons.[24]

We got to the docks and it was a madhouse

A dependent child is fascinated by an Air Force Security Police working dog during the evacuation of dependents from the Philippines.

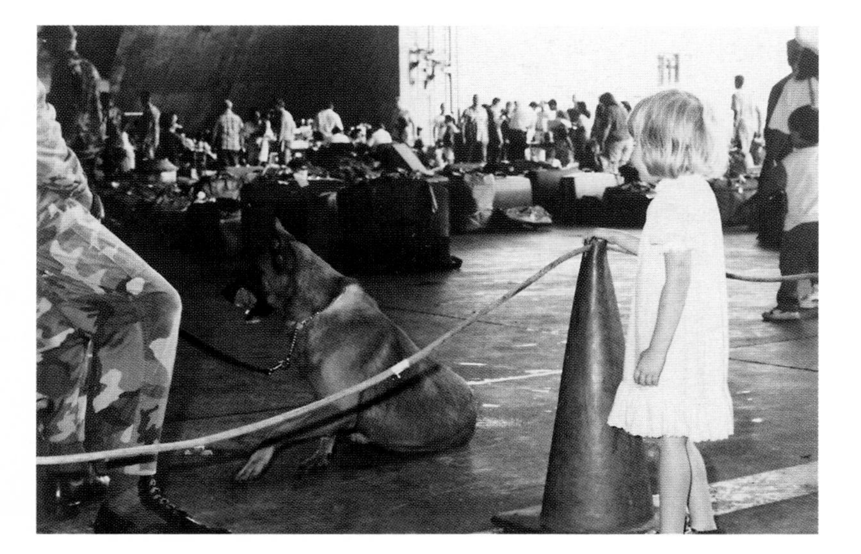

Donna Studer, wife of the Thirteenth Air Force commander, remembered the conditions in the lines. "People were leaving all sorts of things along the line. Suitcases. Everything. It was so hot! The ash was everywhere, and it was like you had been out in the water and then rolled on the beach. It was just everywhere. The lines were just horrible . . . and people were fainting. They set up the gym as an emergency area. People would just leave their suitcases. I was talking to one guy – who was pretty far up the line, and someone came up to him and told him his wife was in the gym and she wants you. He said he couldn't get out of the line or they would lose their place. So I told him I would keep his place in the line. People were desperate to get out of there. They were leaving suitcases full of pictures. There were people sending their Filipina wives back to the states to families they didn't even know. It was really sad to see all of that, and there wasn't anything you could do about it."[25]

The teenaged daughter of another officer continued her story, "We got to the docks and it was a madhouse. There were people everywhere, pets everywhere, and stuff everywhere. All of us were trying to get on a boat and it didn't seem like any of us were going to make it. I'm not sure how it happened, but Mom and I were told to board the USS *Abe Lincoln*. It was a monster of a ship, but the number of people trying to get on was staggering. Even with the ship's size, I didn't think that there was any way that there was going to be enough room for us and the crew on the same ship. Mom and I walked the gangway to get on to the monster and stepped inside . . . the Navy boys were pretty cute and they were so helpful. We got to the entrance to the ship and they immediately took everything from us. They wouldn't even let Mom or me carry our purses. They took it all. They shipped our laundry off and took our stuff to our bunks. They were amazing. There were people everywhere and they all kept their cool. We were sad sights. All of us, not just Mom and me. Nobody had showered or changed their clothes for days. We hadn't had a decent meal in a week and were exhausted. Those guys didn't bat an eye when we asked them where the showers were and didn't laugh when we thanked them profusely for the Kool-Aid. The Navy guys call it bug juice and it was the best thing I had ever had. Every time I drink red Kool-Aid that is over-sugared, I think about those Navy boys. They were so sweet."[26]

Sailing to Cebu

A seaman on the USS *Champlain* remembered, "The most challenging [thing] was having to deal with all the stories about the complete devastation and hardship that the people themselves dealt with. It made all my problems seem very minor. The most rewarding was to see their faces when they reached the ship and [realized they] had survived the evacuation. There was this little girl who slept in my rack* and she said, 'You're Robinson, this is your rack?' And I said, 'No, sweetie, it's yours.' She also made me a picture of the ship that I will always keep. It was one of the best feelings I ever had."[27] The seamen slept wherever they could find excess space in their work areas, passageways and nooks and crannies throughout the ships. The evacuees slept in Navy racks.

The teenaged evacuee continued, "In the center of the ship was an area that they had blocked off for pets. There were a bazillion animals. The dogs that didn't have any kennels were tied to the tiedowns for the airplanes. The navy guys gave them all just enough slack that they could walk a complete circle and get some exercise, but none of them had a long enough chain to reach any of the other dogs. They just kept running in circles barking and snarling as they passed each other. It was pretty funny. They all had pans of water and pans of food. The cats pretty much all had kennels, so they were stacked against the wall of the ship. They were relatively quiet . . . at least ours was! I'm pretty sure that the Navy had kitty litter so that the cats could come out of their cages and go to the bathroom. The cats were all . . . afraid of the dogs, though, and it was quite hard to drag the cats out of their cages and then to get them to pee in front of these snarling dogs who were raring to eat some kitty meat.

"Getting used to the vertical stairways on a Navy aircraft carrier was quite an experience. I tried to slide down them like they do in the movies, but it doesn't really work that way. It's definitely something you want to do right the first time. We found our quarters and they were pretty cool. They were small but cool. We had bunks stacked three high. They were pretty comfortable if you were on the top or middle one, but if you were on the bottom it was pretty rough. You had to lie down on the floor and sort of shimmy in. It looked pretty frustrating. Mom and I were on top and middle bunks and so we made the best of it being so small by telling ourselves we were at camp."[28]

One Navy lieutenant, a 1985 graduate of the Citadel, said, "what better use for the . . . hangar than to be transformed into a barn. Many people have pets that are like members of their family, if our taking care of their animals enable them to relax after their ordeal, then maybe we help them begin the transition back to normal life."[29]

There was a significant problem with motion sickness on the ships that kept the

People and their pets were accommodated in the cavernous lower decks of the ship.

sailors busy insuring the berthing compartments were clean. Many evacuees arrived without shoes and only one set of clothing. Personal hygiene items were unboxed from ships' stores and distributed to the evacuees. Many sailors volunteered to care for children while exhausted parents slept. Curious children wanted to explore the fascinating ships, so sailors escorted them and kept them safe from a myriad of unfamiliar dangers.[30]

During the May planning meeting at Clark, the hospital commander, Col Brian Duffy had expressed concern for the number of pregnant women who might be evacuated. His concerns were well-founded. Six children were born on ships during the trips between Subic Bay and Cebu. On the *Peleliu*, Lt Cmdr Jeff Jensen, MD, delivered the fourth baby to be born on the amphibious assault and landing ship. The baby boy was named John David Decampo. Another baby boy, Abraham Valencia Pineda, would have a lifelong story of how he was the first child to be born aboard the mighty aircraft carrier that gave him his name, the USS *Abraham Lincoln*.[31]

As the newborns arrived at Cebu they were given first priority for airlift to the hospital at Kadena Air Base, Japan, along with other women approaching the end of their pregnancies and a few others who were having medical difficulties. In all, 86 patients, accompanied by 59 family members, were airlifted to Kadena's hospital.[32]

The armada that shuttled evacuees to Cebu from Subic Bay grew to 21 US Navy vessels by

* Bunk.

19 June. Added were the Combat Stores Ship *Spica*, Amphibious Cargo Ship *St Louis*, Frigate *McClusky*, Dock Landing Ship *Comstock*, Amphibious Assault Ship *Peleliu*, the Tank Landing Ships *Bristol County* and *San Bernardino*, Oilers *Hassayampa* and *Ponchatoula*, and Repair Ship *Cape Cod*.[33]

The *Midway* and *Mobile Bay* arrived in Subic June 20. The *Midway* carried helicopters to provide logistical airlift at Subic Bay. She also brought 2,000 cots and enough food for 3,000, as well as diapers and dog food. A team from the Navy hospital at Yokosuka, Japan, augmented Midway's medical department. By June 22, the ships had delivered all military dependents to the Mactan Airfield – a total of more than 17,300. At Clark, General Studer and his unit commanders, realizing the devastated bases could not support the 5,000 troops they had intended to shuttle between Clark and Subic Bay, decided to send 4,500 of them out in the evacuation.[34] The *Midway* made the last run to Mactan with 1,800 evacuees.[35]

Cebu

The *Arkansas* arrived at Cebu first, at 8:30 a.m. on June 17, and deposited its passengers. As soon as the 860 people arrived, problems developed. Rather than keeping the people onboard the ships and processing one aircraft load at a time, all were placed on shore. Processing that many passengers, building baggage pallets, manifesting each aircraft, and keeping the people and pets already manifested segregated from the others proved nearly impossible. There were no waiting facilities on the military side of the field. There was neither cover nor shade. People stood in long lines in temperatures that reached 100-plus degrees. These circumstances predicated a request for housekeeping kits and portable latrines. Local city officials made a few school buses available for transport from the ships to the airfield.[36]

Approaching Cebu, personnel were airlifted by helicopter from the carrier to Mactan.

As the ships arrived in Cebu, they unloaded their passengers using any means available. Helicopters delivered some to the Mactan Airfield. Others traveled via a variety of small and large boats. During the first couple of days of operations at the old Mactan airfield, conditions were difficult. Many people were overcome by the 95-plus-degree heat and the stress of the situation. The ALCE* at Mactan, in response to a request from the 313th Medical Group, asked for air conditioners for the medical tents that were even hotter as they baked in the sun. None could be found, so the 834th CAT at Hickam Air Force Base, Hawaii, shipped a portable aircraft air conditioner. The ALCE also requested port-a-potties to support the operation.[37]

Thirteen C-141 missions flew to Mactan on June 18 and ten more on June 19. Each of these missions carried 143 passengers out of the Philippines. Also, three commercial jet liners, under charter to the USAF, flew into Mactan and extracted 164 passengers each.[38]

However, the flow of aircraft into Mactan inexplicably stopped. Although the person who stopped the flow was not identified, it appears the decision was made for two reasons: there was no fuel available; and aircraft were arriving faster than the ALCE could manifest passengers, put luggage on pallets and deal with hundreds of pets. The research of Anne Bazzell, at the time an airlift historian, revealed that Philippine Airlines denied further fuel to the evacuation effort. This problem was exacerbated when PACOM directed the Navy ships to continue disembarking passengers despite the rapidly building logjam at Mactan.[39]

An agreement was finally reached between an Air Force contracting officer and Philippines Airlines at Mactan for 100,000 gallons of fuel. An additional 300,000 liters was made available from Continental Airlines to accommodate Continental and Key Airlines. After a nearly eight-hour delay the airlift resumed.[40]

The ALCE's earlier request for manpower and equipment proved farsighted in light of the rapidly building throng of evacuees. Two air-transportable medical clinics with the people to man them arrived from Kadena Air Base, Japan, and thousands of MREs arrived, also. From Yokota Air Base, Japan, came one air conditioner, eight pallets of personal comfort items, living facilities, a mobile kitchen and a 50-person team

* *Airlift control element. Coordinated efforts at Mactan.*

of engineers. The 4th Combat Communications Squadron at Andersen Air Force Base, Guam, deployed personnel and equipment to improve communications at the isolated airfield.[41]

Donna Studer recounted her arrival at Mactan, "I don't know how many people were there or how many airplanes had gone out, but it looked like a concentration camp. They kept us in a sort of a penned area. [There were] plenty of portable toilets and plenty of MREs, and plenty of water . . . cots [were] set-up. In fact, I spent one night sitting on a cot talking to one of the volcanologists who was with us on the evacuation.* But people just kept coming in and in and in. People were pulling apart cardboard boxes to sit down on and lie on . . . I didn't write it down in my notes, but I think we were there about 16 to 18 hours. The thing that was worst for me were the mosquitoes . . . they just ate me up. I went to [the medical area], but the poor doctors, they were just overworked. Everyone had done a good job of setting the place up, but they weren't prepared for the numbers of people that were there."[42]

In less than a week, 21,635 evacuees descended on Andersen Air Force Base, Guam, with their few salvaged possessions and family pets

They also were not prepared to deliver babies. Russ Casey, an Air Force master sergeant from Clark was traveling in the evacuation flow with his wife, who was a little over eight months pregnant with their second child. They arrived off-shore Cebu on the USS *Long Beach* and were put aboard a landing craft for a rough, bouncy trip ashore. After processing into the Mactan holding area, they were given priority to depart quickly. However, Mrs. Casey started showing symptoms of an early labor. Rather than board the aircraft for Andersen, Casey tried to find someone to examine his wife at the medical tent. No medical doctor was available, but a veterinarian examined her and confirmed she was in labor. The family was transferred to a local hospital, Cebu Doctors' Hospital, where Mrs. Casey delivered a healthy baby girl. Two days later, the Casey family, now numbering four, returned to Mactan and immediately boarded a flight to Andersen. The only clothing they had for the new-born was a receiving blanket they had packed at the last moment and a diaper purchased at the Filipino hospital. When they arrived at Guam, the wife of a finance officer pulled them out of the lines and took them to her home at Andersen Air Force Base.[43]

* *This was John Ewert, who was rotating back to the US after completing his "tour of duty" monitoring the volcano.*

US Soil At Last!

At Guam, Rear Adm J.B. Perkins, the CINCPAC Representative for Guam, activated the Joint Task Force Marianas to deal with the anticipated influx of evacuees; however, the heaviest load fell upon the people at Andersen Air Force Base. Accommodations were sought all over the island. More than 3,600 were housed at Andersen, in dormitories, tents, and vacant base housing. Another 2,000 spaces were found in other government facilities on Guam.[44]

Processing of evacuees was accomplished in a giant hangar, once used to house and repair B-52s. Evacuees accomplished preliminary clearance of customs and immigration, met with specialists to be accounted for and counseled on assignment and pay and allowance procedures, and were assigned temporary lodging until they could be placed on an eastbound flight. They were authorized an immediate settlement of $2,000 as an advance on damage and loss of possessions.[45]

While the evacuees received care, so did the aircraft. The assorted transports (C-141, C-5, C-130, DC-10, DC-8, B-727, B-747, and L-1011) required attention to make them ready for the long trip back across the Pacific. Members of the 605th Military Airlift Support Squadron (MASS) teamed up with their hosts, the 633d Air Base Wing, for what was to be a test of their skills and endurance. "During the height of the crisis, everyone was on 14-hour days with no days off," related Maj Kathy Runk, 605th Chief of Maintenance. "Our biggest challenge was finding places to park aircraft so we could refuel them. Several aircraft sometimes landed in an hour, so we had to make lots of adjustments, but it worked out."[46]

"Everyone's attitude was great despite the hard work," Major Runk said. "I heard someone on the radio say, 'How are we doing?' Another answered, 'Oh, we must be behind.' Someone else came back with 'No, we're ahead! It's a piece of cake!'"[47]

Practically every unit on Andersen helped out. The 633d Supply Squadron's fuels people pumped 3.8 million gallons of aviation fuel for the thirsty transports. The 633d Transportation Squadron drivers and their Navy and Government of Guam counterparts logged thousands of miles shuttling evacuees between aircraft, the evacuation center and billeting areas. To keep hunger pains to a minimum, the 633d Services Squadron turned 600 loaves of bread, 700 pounds of ham, 250 pounds of cheese, 1,000 pounds of fresh fruit, 6,000 drinks and assorted foods into in-flight meals for the hungry travelers. They also laundered 88,000 towels, 18,000 pillows, 25,000 sheets and 18,000 blankets and pillow cases. Three hundred and twenty-four Andersen residents opened their homes while about 1,100 people logged more than 20,000 hours of volunteer work in the evacuation center.[48]

Hundreds of local volunteers helped carry bags, cared for children, drove cars and buses and lent a willing ear. They heard about the terror and darkness of volcanic eruptions, the exhausting trip, and the kindness and generosity of the officers and men on the Navy ships that ferried them to Mactan.[49]

Pets at Guam – Pets in their cages were stacked in a large holding area at Andersen AFB. Over a thousand animals transited the base.

In less than a week, 21,635 evacuees descended on Andersen Air Force Base, Guam, with their few salvaged possessions and family pets.[50] Pets were airlifted from Mactan in a variety of ways. Those in pet carriers were loaded either into commercial jet cargo areas or put onto pallets for loading onto military transports. However, many pet owners, especially those with dogs, did not have carriers, so an emergency request was made for a shipment of carriers to Guam. Since it was not feasible to slow down the evacuation flow to wait for the pet carriers, some pets were loaded into cardboard boxes and shipped to Guam. Not all survived as some boxes crushed during transit and some unfortunate animals suffocated.

At Guam the pets became an even larger problem than they had been at other points during the evacuation. Guam has strict laws that govern importation of animals. Lengthy quarantine periods and documentation of vaccinations are required. Of course, none of this was possible during the evacuation. Andersen Air Force Base officials negotiated a "work-around" whereby they would quarantine all the animals on the base and gave Guam officials assurances that the animals would be strictly controlled until they were loaded on aircraft for the mainland United States. As aircraft landed at Andersen, a member of the quarantine team boarded the aircraft with a customs inspector. Owners stayed on the aircraft to identify their animals. Once the pet was identified with its owner, the pet was taken to the quarantine area located on a base athletic field. There they were given a medical check and put into the holding area. The carriers were stacked several feet high under tents to protect the animals from the sun. Nearly 1,200 pets transited Andersen during the evacuation.[51]

Also at issue were regulations that prohibit the transport of pets on military aircraft. Transporting the pets in military aircraft had been an issue before the evacuees had ever left Clark. General Studer had discussed the issue with Gen Jimmie V. Adams, the commander of Pacific Air Forces, and Adams supported Studer's desire to allow the pets to travel with the families. Studer's reasoning was that the Clark population already had endured a lengthy hardship, therefore, asking people to leave their pets behind was asking too much.* The issue was decided de facto when the evacuation to Subic Bay was ordered. After the devastation of Black Saturday, a decision to order a mass killing of the pets was unthinkable.[52]

In a letter to her father, a teenaged girl recounted her arrival at Guam. "[The large hangar at Andersen] seemed like a maze of tables and stations. We seemed to have to stop at every table for something or other. I remember finding it amusing that at every stop somebody gave us money, but we really had no use for it. It wasn't like there was anything to buy or hotels for us to stay in. But, we just kept taking it and dreamed about what we would buy when we escaped from the hellish predicament that we were in. Once we had collected all of our money, Kool-Aid, and cookies, we escaped to our little bungalow. It was really nice. I'm pretty sure that there were 3 bedrooms and 2 bathrooms. We all took our first real showers in about two weeks.† The showers we had taken on the ship were navy showers . . . no constant stream of water.

* Most military members were aware that published AF non-combatant emergency evacuation (NEO) plans prohibited evacuation of pets. It was generally assumed pets would be left behind and euthanized.

† Throughout the evacuation, many families banded together in larger groups so they could help each other. The girl is referring to three families that had been traveling together and were living in the same quarters.

You squeezed the button on the hand-held shower head to wet yourself down. Then you lather, then you squeeze and rinse again. It was quite an arduous process. Anyway, we took our first real, no nonsense, no effort, no squeezing involved, stand there and enjoy it showers. It was fabulous. It was very much a cleansing in more ways than one for all of us. I spent my time crying and hoping that everything would end soon. I'm fairly sure that I wasn't the only one who cried in the shower. We had spent a lot of time being strong in the face of adversity, and when we were all alone with our thoughts and our tears it felt pretty good to just let it all out. In fact, it felt great. I said about a million prayers in that 10-minute shower. I thanked God for all of the lives saved and said prayers for those lost. I said prayers of protection for you, Dad, and the others still at Clark, and I said prayers of strength for me and Mom. I also asked for a way home. At that point, I really didn't care how I got home, I just wanted to be there ASAP . . . sometimes God has a sense of humor and he'll give you what you ask for but you have to endure a lot to get it.

When we were all alone with our thoughts and our tears it felt pretty good to just let it all out

"I seem to remember eating real food that night . . . anyway we did get a good night's sleep. The next morning we got up and went to the hangars to figure out what was wrong with the cat and why they wouldn't let us have her. Apparently because of the deaths of other pets on the plane from Cebu, all other pets on those flights had to be quarantined for something like a month. It was ridiculous. I guess that Guam doesn't have any diseased animals and they like to keep it that way. We tried to stress to them that all this hoopla was causing us to get left in the dust when it came to getting a flight out of Guam.[*] We also begged and pleaded for them to let her go just for the fact that [the cat] was old and decrepit anyway and was probably going to die if they kept her in the cage any longer . . . she had been in her cage for three days, I think, maybe longer.

"As the days went on, it seemed like we were never going to get out of Guam . . . there were people everywhere. It seemed like all of the babies were crying . . . at the same time. It was smelly. It was hot. It was loud. Mom and I were in a hangar waiting for a plane for what seemed like forever. I slept on a cardboard box on the floor. I never thought that I would know what it was like to be homeless, but I think we all came pretty close."[53]

The military families who lived at Andersen, following the lead of Navy families at Subic Bay, were generous, and many opened their homes to the beleaguered evacuees under a base program called "Adopt a Family." One family did more than their fair share. SSgt Dal and Janice Whelpley and their three children housed more than 17 families during the days evacuees were travelling through the base. Mrs. Whelpley said, "The full impact finally hit me when I walked into the big hangar, and I saw all those people with nowhere to go. My husband looked at me, and I looked at him. He went to work at the billeting table, and I went to Family Services to bring home our first families."

Mrs. Whelpley continued, "I picked up four families the first day, and I've actually lost count since then. I'd pick up one family from the hangar, return home to settle them in, and go back to pick up more. We were just trying to be as helpful as possible. It got pretty hectic around here. One time we had at least four children living here named Jessica. We just started calling them Jessica One, Jessica Two and so on. It was pretty funny.

"One [evacuee], Elita, even managed to look beyond her troubles to help some of the other evacuees. Even though she had lost everything, she didn't have a bad outlook on life. It was really inspiring for me because no matter how down everybody else was, she kind of brought them up."

Mrs. Whelpley and her friend, Denise Bodam, like many of the host families did their best to make the evacuees comfortable. "We took them around to all the places that they needed to go. We made sure that they had everything they needed. We bought the food they wanted, so that they could prepare the kinds of food they liked. I think they were pretty happy and satisfied with their stay."

When a reporter asked Mrs. Whelpley if she would do it all again, she responded, "In a heartbeat. Except this time, I'd take in more."[54]

Another Andersen wife, Carolynn Farrell, related how she selected families to take to her home. "Some women, exhausted after carrying a new-born, a toddler and two very heavy bags containing all that they now owned,[†] would sit

* *Although the decision had been made to allow pets on military transports, there was still confusion at Andersen. Initially, evacuees travelling with pets were told they would have to wait until commercial aircraft, which could carry pets, were available. Ultimately, families and their pets were loaded on both types of aircraft to accelerate the flow back to the mainland US.*

† *Rumors ran rampant concerning the conditions at Clark. Many evacuees believed their homes had been destroyed. Most could not foresee a circumstance where their belongings could be packed and sent to them.*

down and cry. Others, too tired to sit and think, would stand and cry – still clutching a confused toddler and sleeping baby. These are the women that I took home with me."[55]

The large hangars that were used to process the evacuees were crowded as evacuees went from station to station manned with volunteers from the USAF, USN, USA, and USMC. They first processed through customs officials who searched luggage and cleared it for entry. The luggage was then set aside so the evacuees would not have to lug it through the long lines. A first aid station and medical desk, where doctors, nurses and volunteers worked 12-hour shifts, offered assistance to those with health issues. Base chaplains and evacuee chaplains worked at a station and patrolled the processing areas offering assistance and advice where needed. Another station offered day-care services and on-the-spot baby-sitting for those who wanted it. A Red Cross desk offered message service so evacuees could send messages to their families in the United States. An international telephone company provided free telephone calls, one-per-family, to any location in America. The most prized possession of the long lines, though, was a blue piece of paper that established the holder's position in the queue waiting for a transport to take them the rest of the way home.[56]

Evacuees board a ship taking them to safety.

Home

Planeload after planeload departed Andersen for a refueling stop at Hickam Air Force Base, Hawaii, and their final destinations at McChord, Travis, and Norton Air Force Bases. As evacuees made their way across the Pacific, repatriation centers were set up at the bases. These reception centers provided comprehensive services, including child care, medical care, escorts, financial and legal assistance, airline flight information, interpreters, food centers, pet areas, hygiene supplies and counselors. The Red Cross, United Services Organization (USO), and Salvation Army aided the evacuees. At Hickam, the Red Cross provided snacks and drinks, while hundreds of volunteers worked round-the-clock carrying bags, babies, and pets, allowing exhausted evacuees to rest before the final leg to the United States.[57]

After arriving at a stateside port, the evacuees again went through the long processing lines. An escort was assigned to each family as they got off the airplane. The escorts, military personnel who volunteered their time, shepherded the evacuees through the myriad stations waiting to process them into the United States.

The first step was to clear immigration and customs. By this stage the evacuees had been processed several times, so any illegal travelers had been picked out. Master Sergeant Casey, whose daughter had been born three days earlier at Mactan, was frustrated and angry to find an immigration officer who demanded a passport for the newborn. A supervisor quickly settled the dispute, and the Casey family, intact, entered the United States.[58]

One of the last steps for the evacuees was to be issued a free airline ticket to a location of their choice on the mainland United States. Kathy Wallace, a scheduled airline ticket office (SATO)* employee from Nellis Air Force Base, Nevada, had volunteered to work at McChord Air Force Base to augment its swamped travel officers. Less than two hours after she heard about the need for volunteers, Mrs. Wallace was on an airplane to Seattle, where she rented a car and drove to McChord. She recalled, "There were tons of workers already there when I arrived, but they had been working around the clock and were exhausted. Volunteers from the base were bringing in food and drinks to keep them going. We had a whole assembly line. The escorts would pick them up off the plane as soon as they came in and fill out a slip that said where they wanted to go. Most of these planes that came in were full of women and children. We would then book them on airlines to where they wanted to go out of Seattle airport.

"They were all exhausted. A play room was set up for the children where they could lay on mats and watch television. Red Cross had tents set up outside. Some of [the evacuees] only came with a garbage bag full of clothes. A lot of them just came with the shirt on their back. They could pick up clothes and toys at the Red Cross tent. There were [many] pregnant women, plus the ones who had babies on the ships. I've

* SATO operates on most military bases to provide commercial air travel for service people and dependents.

never seen so many women and children. Some of the women could hardly speak English, plus [some] parents were with them.* Some of them were going to families whose names they did not know and whom they had never even met. It was just overwhelming."[59]

No matter which base the evacuees arrived at, one feeling was universal – relief. The teenager continued in a letter to her father, "When we landed [at Travis] we were greeted by AF personnel dressed in their mess dresses with blankets. There were cameras everywhere and reporters asking questions. I've never experienced anything like it. These people were so happy to see us, and I'm sure that we were even happier to see them. I kissed the runway and said a prayer for me and Mom's safe arrival on solid ground that wasn't moving, breaking up, or exploding. We made our way into the airport and once again began the ritual of collecting money and cookies. The people at McChord were even more prepared for us. They had everything known to man. I bought some magazines and waited while Mom found us a flight to [grandmother's].

"I learned some patience and some other things, too. I learned that mothers are tough as nails, people really can come together when they have to, there's no place like home, singing can calm any scary situation, the shower is the best place to cry, not to be afraid of God when he's trying to get your attention . . . he just wants you to listen for a minute, and that I can sleep anytime, anywhere, and in any position. That's a good skill to have when confronted by a big scary mountain with nothing better to do but blow up and ruin my party."[60]

Nearly 20,000† evacuees spread out to homes across the United States to "Safe Haven" locations. Gradually the military men and women that had been held at Subic Bay caught up with their families. Families of service men and women still at Clark had to wait longer. Everyone wondered about the fate of Clark, but they were most concerned about their personal possessions that had been left behind on June 10, 1991.

*It was not unusual for an American serviceman, who had married a Filipina, to adopt some members of his wife's family. Extended families that take care of each other are an important part of Philippines' culture.

† Some had left the evacuation flow at Guam and Hawaii to join families there.

NOTES TO CHAPTER 7

1 Interview, Donna Studer with author, January 5, 1999 [hereafter cited as Studer, Donna interview].

2 Conversations, Amy Anderegg with author, September 1998-February 1999 [hereafter cited as Anderegg, Amy conversations].

3 JJTF-FV SITREP, 101100Z Jun 91.

4 End of Tour Report, Maj Gen William A. Studer, 5 January 1990-1 December 1991, p 2.

5 Studer interview.

6 JTF-FV SITREP, 121100Z Jun 91.

7 Studer interview.

8 JTF-FV SITREP, 121000Z Jun 91.

9 JTF-FV SITREP, 131300Z Jun 91.

10 JTF-FV SITREP, 131300Z Jun 91.

11 JTF-FV SITREP, 141645Z Jun 91.

12 Bazzell, Anne M., Joint Task Force-Fiery Vigil, 8 June-1 July 1991, 834th Airlift Division, Hickam AFB, Hawaii, 96853-5000, 1 April 1992 [hereafter cited as Bazzell].

13 Ibid., p 10.

14 Ibid., p 13.

15 Ibid., p 14.

16 News article, "Honolulu Star Bulletin," June 26, 1991, p B-3.

17 Goodman interview.

18 Bazzell, p 24.

19 Goodman interview.

20 Conversations, Jean Anderegg with author, September 1998-February 1999 [hereafter cited as Anderegg, Jean conversations].

21 Anderegg, Amy conversations.

22 JTF-FIERY VIGIL SITREP, 181415Z Jun 91.

23 Anderegg, Jean conversations.

24 Studer, Donna interview.

25 Ibid.

26 Anderegg, Amy conversations.

27 Robinson, Daniel K., E-6 USN, Msg from USS Champlain to NAVPACEN, San Diego, 070834 Jul 91.

28 Anderegg, Amy conversations.

29 Taylor, Barry R., Lt USN, interview for NAVPACEN.

30 Williams, Derrick R., E-6 USN, interview for NAVPACEN.

31 Message, USS Peleliu to COMSEVENTHFLT, 210233Z Jun 91.

32 Bazzell, p 17.

33 Ibid., p 21.

34 AMEMB SITREP 32, Subj: Mt Pinatubo, 181658Z Jun 91.

35 Bazzell, p 23.

36 Ibid., p 15.

37Ibid., p 17.

38 Ibid.

39 Ibid.

40 Ibid.

41Ibid., p 18.

42 Studer, Donna interview.

43 Interview, MSgt Russell W. Casey, with author, January 16, 1999 [hereafter cited as Casey interview].

44 Bazzell, p 16.

45 Ibid.p 19.

46 Ibid.

47 Ibid.

48 Ibid., p 21.

49 Ibid.

50 Ibid., p 25.

51 Tropic Topics, official newspaper of Andersen AFB, Guam, date unknown, ca. June 17, 1991 [hereafter cited as Tropic Topics].

52 Studer, interview.

53 Letter, Amy Anderegg to her father, n.d.

54 Tropic Topics.

55 Ibid.

56 Ibid.

57 Bazzell, p 22.

58 Casey interview.

59 Interview, Kathleen M. Wallace with author, January 16, 1999.

60 Anderegg, Amy conversations.

Aerial view of Clark Air Base flight line

3rd Combat Support Group Headquarters

View from the control tower

Clark Air Base control tower

Aerial view of the impending eruption

The wait is over

Fiery warning

Acacia Trees-Before...
Magnificent acacia trees lined many streets in the housing areas and other large areas of the base.

Acacia Trees-After...
Wet ash fall devastated the acacia trees. Incredibly, all had fully recovered by 1998.

A picturesque 1.2 mile-wide lake in 1998 was once Pinatubo's fiery crater.

All color disappears as the ash decends

Abacan walking bridge

Ash driven by the typhoon

Officers' Club parking lot

Chambers Hall in 1998

A frightened child receives comfort

An original picture of the Fort Stotsenburg entrance.

The Parade Field in 1998. Note the date 1902 barely visible after the ash fall.

Chapter 8 PACKING OUT

In Washington, DC, civilian and military leaders of the Air Force discussed the fate of Clark Air Base. On the other side of the world, the Ash Warriors attacked the formidable task of packing the household goods trapped in 3,000 military family homes on the base, several hundred homes off the base, and a like number of dormitory rooms in the base's barracks. The official nomenclature for these possessions is household goods, but those impersonal words do not reflect the emotional value those possessions held for the Ash Warriors. Hidden behind a nearly impassable tangle of branches and under a blanket of mud and ash, which had hardened into crunchy concrete, were children's first pairs of shoes, wedding photos, christening gowns, grandmothers' china sets, high school diplomas and other treasures of the heart. Although such possessions are valued by every family, they held an even more emotional significance to a military family that moved around the world during years of military service. They represented ties to loved ones, neighbors and places that had not been seen in a long time and might never be seen again.

As soon as water, food, and shelter were secured at Clark, the Ash Warriors turned their efforts to saving as much as possible. The problems were many. The volcano continued to erupt huge, swirling, black clouds of ash that frequently drifted across the base. Sometimes it was so thick that visibility was near zero and all work outside had to stop. Branches from the once-beautiful acacia trees, whose expanses had shaded the residential areas, clogged streets, driveways, and entryways to the homes. The ash on the streets made driving very difficult for normal vehicles and impossible for moving vans. However, the Ash Warriors' worst enemy was the approaching southwest monsoon that would bring the rainy season.

During the eruptions, falling ash and rain filled the base drainage systems and sewers. Essentially the broad expanses of the base were made smooth; there were no ditches or curbs to direct the water in predictable directions. The result was an effect called "sheeting." Whenever rain fell, it flowed in huge sheets across the ash until it eventually eroded new patterns of flow. However, these patterns did not remain constant. The next rainfall would wash new patterns into newer ones, and the surging rain waters might capriciously change directions.

The surging rain water during the eruptions on Black Saturday had left plenty of visible evidence of their potential. Rushing water and mud had blasted through the doors of several homes as well as large, seemingly stronger structures on the base. The volcanologists advised the Clark CAT to move first the lowest areas on the base. Another scientist, Dick Janda, joined the team shortly after Black Saturday. Janda was an expert on lahars, the mud flows that inevitably follow an eruption like Pinatubo's. Janda worked closely with planners from the 3d Transportation Squadron, commanded by Lt Col Doug Steward (Hotwheels), to divide the base into sectors. The highest priority sectors were the ones that seemed most vulnerable to mud flows by virtue of their low altitude or proximity to paths that water had already eroded through the ash.

Usually, the rainy season came to Clark in July. Every ounce of effort the Ash Warriors could muster was aimed at one goal: beat the rain. Eagle set a goal of 100 homes per day. If the transportation squadron packed out 100 per day, it would take 30 days to complete the

The first priority of clean up was to shovel the roofs clear of tons of ash to make safe the work and living areas.

effort. In normal times, the base had been able to pack out 20 or 30 homes a day. Hotwheels recalled the goal as "impossible." He understood well the effort moving that many homes a day would take. For example, an overseas packing crate held 1,000 pounds and was constructed of 8 sheets of 4' x 8' plywood, and an average home held 6,000 pounds of goods. Therefore, each home required 48 sheets of plywood, so 100 homes required 4,800 sheets of plywood per day.

The local moving companies that the base contracted to move household goods had suffered extensive damage to their facilities as well. Workers employed by the moving companies were living in very bad conditions in Angeles, and their families were suffering from dangerous shortages of food and water. However, Hotwheels was determined to make every effort to pack as fast as possible in order to beat the rains, and he had just the person who could make it happen.

TSgt Renee Longstreet, NCOIC of the Outbound Household Goods section of the transportation squadron spearheaded the pack-out effort of the Ash Warriors' personal possessions. Longstreet had been assigned to Clark only three months earlier on an unaccompanied tour without her husband and 15-month old son. She had extensive experience in moving household goods, but her work area was unusable.

She recalled, "When we saw our building was destroyed, I took a team of the people in. It was not a safe area. We went in understanding that we had to be very careful so we didn't disturb things so something came down on us. We wanted to go in and get as many documents, salvageable documents, as we could and equipment, because we could not work out of the building at all. We relocated to the motor pool and got everything set up there."[1]

A more difficult problem for Longstreet was slow and often-changing direction from base leadership. Initially, she was told to move any household goods as fast as she could, and the first people at her desk were some base school teachers who lived off-base. She filled out their applications, but before she could get movers to the homes, she was told to move only homes on-base. The area to which she was directed, in her recollection, was an officers' housing area. However, before any significant packing was done there, the CAT finally gave her group commander a third and final plan.[2]

Hidden behind a nearly impassable tangle of branches and under a blanket of mud and ash, which had hardened into crunchy concrete, were children's first pairs of shoes, wedding photos, christening gowns, grandmothers' china sets, high school diplomas and other treasures of the heart

The final plan, which Longstreet implemented, was designed around Janda's advice on where he thought the mud would flow next. Low areas were to be packed first; higher areas last. The first area attacked was 250 homes near the Sapang Bato entrance to the base. The CAT, following Janda's advice, selected the area because it was the lowest on the base and was only a stone's throw from a large branch of the river that ran just to the south. Engineers plunged into the area with dozens of local workers, chain saws, bulldozers, and road graders. The "Dirt Boys of CE"* who cleared the streets and driveways was led by CMSgt Dan Anderson. His 65-person team, augmented by a Red Horse† team from Osan Air Base, Korea, worked exhausting hours opening access to the homes.[3]

As soon as the first streets were cleared, several teams entered the area. Each team consisted of a legal representative,‡ a customs inspector and a house monitor. Whenever possible, the house monitor was from the unit to which the occupant was assigned. The team's first problem was to confirm they were in the correct house. Sometimes, the base housing office was able to provide a key and the owner's name; however, the housing office and its computerized records had suffered considerable water and mildew damage so the records were incomplete. At times, the team had to search personal records to identify the occupant. Since most occupants had taken their keys with them, locksmiths had to open, repair, and re-key nearly all the 3,000 houses.[4] "Picking a lock isn't as easy as TV portrays it," said Sgt Kenneth Schetroma, a locksmith, "it sometimes takes hammers."[5]

* Civil Engineering.

† Air Force combat engineering team.

‡ Lawyers from legal offices at other PACAF bases and US bases came on temporary duty (TDY) to supplement the teams.

Once the team was positive they knew who owned the goods, they did a quick walk-through of the house to determine what could be shipped. The first teams in the houses were shocked sometimes at what they found. Although most homes seemed undamaged when viewed from the outside, the inside was often a different story. The homes had been without air conditioning or any fresh airflow for at least two weeks (several weeks for the homes packed later) and the ever-present mildew had already started to turn walls and ceilings black. Food left in kitchens rotted on countertops and trash-cans. Large fish tanks were dark green with algae, and the putrid corpses of once-beautiful tropical fish floated on their surfaces. The smell in many homes was nearly overpowering, but the worst of it were the refrigerators and the occasional dead family pet.[6]

The teams quickly learned that the refrigerators were a loss. At first they tried to empty them and clean them up for shipment, but the task was impossible. The smell of rotting food made cleaning the appliances a task few could stomach. Ultimately, teams sealed the doors with duct tape and moved them out of the houses where they were abandoned.

Many pets were left behind. Some dogs were left tied outside, and those were released by security patrols to fend for themselves. Some dogs were discovered in homes by security patrols if the animal happened to bark when a patrol was near. Concerned Ash Warrior patrolmen also released or adopted those dogs by either finding the key at the housing office or by just kicking in the door.* However, cats and other small animals left behind seldom survived.

Although no homes on the base collapsed under the ash, many roofs sprang leaks and muddy water seeped under doors and window ledges. Carpets, draperies and stuffed furniture were ruined in many homes.

The inspection and packing teams tried their best to prepare the goods for packing as though it were their own things being packed.[7] They washed dirty dishes from the sinks and threw out spoiled foods. They searched the homes for valuables and boxed jewelry, photos and important documents then mailed the boxes directly to the owners. They filled out an inventory of everything in the house that was shippable and noted on the inventories things that were not shipped such as refrigerators, bedding, curtains, clothing, and carpets ruined by the volcano.† The legal office representatives, organized by SMSgt Jeannie Shaw,[8] confirmed documentation, and customs inspectors insured that no illegal items were packed for shipping.

As soon as the inspection team finished assessing the house, the local moving company assigned to move the goods came in and packed the items for shipping. Someone from the team stayed with the movers during the pack-out to discourage theft. Although there were certainly instances where movers stole items from the houses they packed out, efforts were made to discourage thieves from prospering at the expense of displaced Americans. The security police set up sting teams that "planted" valuables in homes then arrested movers who made a grab for the items.

Capt Erin Gannon, of the 6200th Tactical Fighter Training Group, was on temporary duty in Alaska helping to execute PACAF's first Cope Thunder North exercise when the volcano devastated Clark. Along with her squadron commander and one other member, she returned to Clark to close down the Cope Thunder operation and to help ship household goods and vehicles that belonged to her squadron mates.

Captain Gannon recalled, "[The effort] was really well organized at the beginning, and once people got into the mode of understanding how we needed to function as a team it worked out well. We would have an address, name of the person, and the team would congregate outside that home when it was time to do that house. Usually we had two or three homes a day that we were supposed to go get shipped out. The team would meet and we would wait till the whole team was there and then we'd all go in at once.

"The weather was really hot. You're in your BDUs, and it's like you are standing in

Packers loaded up 3,000 homes in less than two months in a mad rush to beat the rainy season.

* *Rumors circulated that patrolmen shot these stray animals, but no record of such shootings has been found. The security commander said that no pets in their homes were shot by his forces.*

† *Such documentation made it easier for families to make claims for reimbursement to the government or insurance companies. The legal office at Clark inspected over 5,000 homes off- and on-base and paid over $650,000 for more than 840 claims. (Source: Clark Air Base Closure Report, signed by Col Bruce Freeman, 3 TFW commander, November 26, 1991, PACAF Command Historian's Office)*

snow banks. Literally like having to walk through two feet of snow to get up to these houses because of the ash. Most of these areas, the team would gather outside, and most of these areas it would be fairly easy access, [but in] others we would have shovels, and we would have to dig our way, a path, to get into the house and have an easy access path to get their household goods out. So, there were a lot of those things going on, and I remember thinking, it's like it snowed here and we're all standing around outside, and it's either sleeting, it's raining with the dust and ash coming down on you, just feeling like we were in some other kind of atmosphere. It was a real dichotomy because, visually it felt like snow, but it was boiling hot.

There was massive destruction of some facilities. Most buildings weren't designed to withstand tons of ash, drenched by typhoon rain.

"One of the worst things about going into these homes was that the refrigerators were full of food and some of these houses, this was a month after the refrigerator had sat there, and so the homes really, really smelled bad. So . . . the first thing on the agenda was for the team to go into the house and open the windows. I had a standard thing, I'd go in and tape up the refrigerator with duct tape, because your [normal reaction] when it's boiling hot outside is to walk up to the refrigerator and open it [without thinking].

"You always felt you were intruding very much on people's personal lives because normally when a military person gets ready to move . . .you have things organized for the move. People know how to move in the military. Well, [the homes] were just like after a normal day . . . because people just evacuated, the homes weren't ready. So, we had to make decisions. Is this pile of newspapers important to these people, or is it something that I can allow to be trashed? We ended up shipping things that, in my mind, I would never ship if it was my home and I was getting ready to move. You just didn't know if a stack of papers, pile of books or a thing of magazines was important to people. I personally felt I had to try to respect people's privacy and try not to look through their things.

Virtually everything on the base was packed - homes, offices and dormitories.

"At the beginning, the first couple of days, I wished I was [moving people I didn't know], but after that I felt very good because people from the squadron understood I was doing for

them what they would have done for me. [Sometimes], they would call and say, 'could you go in my house and get this special document and mail it to me?' We were doing special things like that for people which I felt good about. Some people . . . didn't understand what we were going through and they were asking ridiculous things, 'Can you send my kids' winter coats?' Other people had things they really needed, a set of car keys, or some important document, a wedding album.

"I felt really good, whether it was a squadron member or someone else, that they were getting their belongings, and we were able to take care of things and were paying attention as they would be if they were there. We'd find their stereo boxes and put things together. I'd take down the curtains and fold them nice and do things.

"Most people, virtually everyone, left their homes not even thinking they'd be gone for a couple days. I was not there to talk to people as they were evacuating, to see what people's emotions were, or what they really thought, but from the state of their homes, across the board, they all thought they were coming back. Everyone left their garbage, like we'll be back in a couple of hours. One guy left his ID card on the counter. That really floored me!"*

Not everything the teams found in the homes made them feel good. Gannon continued, "Maybe after a day or two [security police and OSI agents] were added to the teams because people were finding things in [a few] household

* Nearly any business a military member does with a military institution requires the member to present a valid military identification card. Recruits quickly learn to always have it in their possession.

goods that were really at issue. I mean there were drugs found, there was pornography.* If you find drugs in someone's house . . . we would still pack up those people's goods, [but] their goods would be "red-lined," which would mean that the member would have to report to TMO† on the receiving end. And then that member's commander would be dealing with him from a legal aspect in terms of what type of contraband or illegal items were found in the house. But the security police got kind of rambunctious in terms of . . . I remember one incident where people had pictures of their family, stacks of photos, and they had one picture of a nude little baby, and [the police] said it was child pornography. And it was those people's kid in among the Christmas pictures, and it was not in my mind pornography . . . they were just really zealous about it. And we found a shoe box in somebody's house that was full of classified [information]."‡[9]

A week into the packing effort, only 30 homes a day were being packed. Steward and other commanders huddled with logisticians and transporters. They determined that local workers were losing time traveling to and from work. A deal was struck with housing officials, security police, services, transportation, and the moving companies. They opened base dormitories in which the workers were allowed to live and provided them with meals. The scheme resulted in an immediate increase of at least 25 percent more work time for the packers.

Another decision that greatly accelerated the household goods pack out was one to consolidate the sea-van** loading on the former Cope Thunder ramp.†† A very large area that had accommodated Cope Thunder aircraft for more than 15 years, its main feature was an enormous open-ended hangar with an arched roof. The odd-looking hangar was nicknamed the "wind tunnel" and was a landmark familiar to everyone who flew in Cope Thunder exercises. The entire ramp and the wind tunnel were turned into a staging area for the movers. Trucks brought household goods that had been loaded into wooden crates to the Cope Thunder ramp where they were loaded into sea-vans and sealed. They were sealed to prevent pilferage during transit to the United States.

The packing teams slowly accelerated the rate to 70 houses per day, but that rate held there for three weeks. Despite the best efforts of Longstreet and her team, the goal of moving 100 houses a day was not achieved until over 75 percent of the pack-outs were complete. The rainy season and its attendant typhoon arrivals, combined with frequent blasts of ash from the volcano, slowed the effort. Crate-building materials, which were supplied by the moving contractors, nearly ran out halfway through the campaign as well. Eagle would later say, "It may have been a goal, and [some] said [it was an] unrealistic goal, but we shot for it, and surprising as those things are, you give somebody something to shoot for, and they will achieve it."[10]

Security of the household goods was a serious concern. Sergeant Longstreet, in coordination with headquarters at both Hickam and the Pentagon, elected to ship all household goods back to the mainland United States, even the ones that were ultimately destined for delivery to bases in Guam, Japan, Korea, and other Pacific bases. Her decision was based on one factor. Goods that were shipped to Pacific bases were not sealed in sea-van containers when they left the base. However, goods that went back to the States were sealed into their containers at Clark and stayed in the containers until they arrived in Oakland, California. Of course, it was more expensive and took longer for the goods to go all the way to California and then back to Japan or

The goal was 100 houses packed out in a day.

* Possession of drugs by a military member is illegal. Possession of pornography is not, but it may not be shipped in household goods.

† Traffic Management Office.

‡ Confidential or secret material. Having such material in one's home is a serious security breach.

** A tractor-trailer body that can be stacked until time for transport at which time it is lifted by a crane onto a tractor-trailer frame and driven away. At the port it is then lifted onto a ship. Technically, they are sea/land vans since they can be carried either on trucks or ships. Ash Warriors sometimes referred to them as sea vans.

†† A ramp is Air Force terminology for an aircraft parking area.

Guam, but in her view the expense was worth the trade-off for security.[11] She recalled, "It was not cost effective, but I felt the people went through enough, and even though they had to wait longer to get it, at least they got something. I had people call me up on the phone who were upset, they were irate. I felt for them but I said, 'OK, I had to make a decision. I felt you would have liked to have gotten whatever was left of your property, here, than not get anything at all by letting it get pilfered.'"[12] Ash Warriors such as Japan-bound Russ Casey, whose wife had delivered a daughter in Cebu, had to wait several months before they got their personal possessions, but they received everything that was shipped to them.[13]

Rains from the southwest monsoon and typhoons continued to hit the base throughout the packing effort. Roads that the Dirt Boys from CE had recently plowed filled again with muddy ash and became impassable.[14] Whenever ash from a recent eruption mixed with rain, it clogged transformers and shorted power lines. The plan to move low-lying areas first was justified when floods from heavy rain swept through the area three days after the Sapang Bato area saw the last household goods removed. Many homes in that sector were damaged and flooded.

On July 17, 1991 a joint US-Philippines announcement stated that Clark Air Base would be closed

After the household goods from base housing were packed up, Sergeant Longstreet and her section turned to the task of packing the homes that were off-base. The decision to pack the off-base homes last had been a difficult one. When the pack-out first started, Eagle had serious reservations about the safety and security of starting a large-scale effort to pack-out the off-base houses.[15] The rivers and bridges of Angeles were already ravaged by lahars with the threat of more to come. He feared that teams might be trapped off the base or worse caught by lahars trying to return to the base. Additionally, security forces were stretched to their limits guarding the base housing and government property. If he were to send security forces with the movers, he would diminish security within the base perimeter.[16] He elected to pack the on-base property first, hoping that the eruptions would subside and a better understanding of the lahar patterns would emerge before the moving teams moved outside the base perimeter. The decision abandoned some off-base residences to looters.

The aerial port in 1998.

Property in some other off-base homes fell victim to landlords. Ash Warriors in the United States depended on a friend or squadron member still at Clark to check on their off-base homes. As the on-base pack-out progressed more and more people were reporting that some off-base homes were entirely empty – bare floors and bare walls. After some brief investigation by Col Royle Carrington, Clark's head lawyer, it was determined that landlords were emptying their former tenants' homes and holding the goods as collateral for rent that the landlords claimed was owed them. Under the circumstances, there was no legal way to deny the landlords' claims. A team from the legal office was dispatched into Angeles to "pay off" the landlords and rescue the Americans' property so it could be shipped home.

GOVERNMENT PROPERTY

On July 17, 1991 a joint US-Philippines announcement stated that Clark Air Base would be closed.* The announcement lifted a great burden from everyone on the devastated base. The mission was clear: move out. The Ash Warriors began the task of salvaging government property. Millions of dollars of equipment was trapped in several supply warehouses that had collapsed. The regional medical center held tons of valuable examination and treatment equipment. All of the equipment for the 3d Tactical Fighter Wing's F-4s and the 353d Special Operations Wing's aircraft remained. Hundreds of tons of munitions were in the MSA storage igloos. Over three million gallons of jet fuel and another million gallons of diesel fuel and gasoline were on the base.[17]

One of the most important units on the base had been the 624th Military Airlift Support

* *Details of this announcement are discussed in Chapter 9.*

Group, commanded by Col Al Schweizer. Before the eruptions, the 624th was responsible for all en route support for Military Airlift Command (MAC)* flight crews and aircraft transiting Clark. They also executed the airlift schedule through the Far East and Indian Ocean. More than 700 people were assigned to the unit's three squadrons. As MAC's representative on the base, Schweizer and his troops were tasked to remove all MAC equipment they could find, to include digging it out, and preparing it for shipment to other MAC bases in the area. They processed fork lifts, K-loaders,† tractors, spare C-5 engines, aerospace ground equipment, tools, supplies and several pallet loads of F-16 spare parts that were stranded in collapsed warehouses. Their equipment was doled out to bases in Guam, Japan, Korea, Hawaii, Diego Garcia, and Singapore.[18]

At the base hospital, one of PACAF's large regional medical centers, workers packed expensive medical equipment as they sloshed through muddy ash which had flooded all the floors through huge leaks in the roof. By mid July the effort was going swimmingly and the CAT-scan equipment was moved to Subic Bay.[19] Due to lack of power to operate the hospital's elevators, many bulky and expensive medical items were initially stranded on the second floor. By using two 4,000-pound forklifts and a borrowed crane they were able to salvage these items. Three weeks later all the radiology equipment was shipped, and by early September the huge building was empty of valuable, salvageable equipment except for an "angio suite" that was shipped in the next two weeks.[20] Medical and dental records, which were removed before the evacuation and taken to Subic Bay in a school bus, arrived at Kadena Air Base, Japan, where they were held until families called for them. Ultimately, they salvaged over $23 million of medical and dental equipment.[21] While they packed they still cared for the Ash Warriors. The day after the volcano erupted, the 657th Tactical Hospital, commanded by Lt Col (Dr.) Charles B. Green, erected a mobile facility and started seeing patients. The 23-person unit worked 12-hour shifts and saw more than 100 patients a day for everything from broken bones to aspirin dispensing. They also tested water and food supplies.[22]

The unenviable task of accounting for all government property on the base fell to Col Randy Miller's 3d Supply Squadron. Essentially, their challenge fell into two general categories: accepting the turn-in of supplies from units across the base and recovery of supplies in the giant, collapsed supply warehouses. The task was formidable, according to a staff officer at PACAF, who recalled his first telephone contact with Clark's deputy supply commander, "he described destruction that was hard to imagine – half of the supply warehouses had collapsed roofs, there was no electricity or running water, buildings were down all over the base, and the main base computer had flooded and water was coming out of the disk drives. While he tried to describe the devastation, his words did not convey the true magnitude of the disaster as much as his voice. I have known him since 1983, and he has always been one of those upbeat 'make light of any adversity' type of guys. I'd never known him to be really down until I spoke with him on the phone that day. His sense of defeat came through loud and clear, and I knew then that we had our work cut out for us."[23]

Initially, supply people worked at the supply squadron complex. However, the main administrative building was badly damaged. So, they relocated their operation to the 3d Combat Repair Squadron building on the flightline, which was undamaged. The building had small loading docks and open floor space to accommodate equipment turn-ins. As the recovery and turn-in of equipment progressed, though, the volume of gear increased dramatically, and the operation was moved to the aerial port‡ where they established a packing and crating operation as well. Miller recalled, "It was great – items could be turned in, initially accounted for, documented for shipment, packed, and loaded at one building."[24]

Blowing ash made packout difficult but not impossible.

The most dangerous job was recovering millions of dollars of supplies from the collapsed warehouses. Engineers positioned a large crane beside the first collapsed building, then

* MAC operated most of the USAF's transport aircraft.

† A large, expensive piece of equipment that loads cargo on transport aircraft.

‡ A large area on the flightline where cargo was prepared to be loaded on transport jets, and where downloaded cargo was prepared for base distribution.

slowly extracted the tangled roof panels and beams one at a time. Incredibly, they discovered that almost all the equipment underneath the mess was undamaged. Much of the equipment, which included jet engines, spare parts, precision tools and test equipment, was packed in individual containers or heavily wrapped in plastic, all of which served to protect the mummified items from ash and rain.

The equipment in the warehouses belonged to many base units, and as it was recovered they returned the items to the Air Force supply system. One of the warehouses, the one solely dedicated to the supply squadron, had contained $211.4 million in equipment before Black Saturday. The reclamation effort recovered $190.9 million of it.[25]

THE POVS

Thousands of abandoned privately owned vehicles were scattered across the base. Many were "Clarkmobiles," vehicles that were passed from airman to airman over the years as new arrivals replaced those ending their tours. Most Clarkmobiles were not worth much money, but some were valuable. All the vehicles were dual-registered; they carried Philippines license plates, and they had stickers on windshields or bumpers issued by base security police.

Shovel first - then pack!

The transportation squadron and legal office undertook the arduous task of identifying the POVs and moving them to a large holding area near the flightline. Many of the vehicles were blocked from access by the large piles of ash pushed against their sides as road graders plowed the streets. So, they often had to be shoveled out by hand in scenes reminiscent of the days following a blizzard in New England. Once the POVs were free of ash and branches, they were towed to the holding area. The primary concern was to get them off the streets, especially in the housing areas, so the household goods teams and movers could get the homes packed.

Many evacuees telephoned friends or squadron mates at Clark and asked them to ship the evacuee's vehicle to the United States for them. Gannon recalled helping out in one such instance, "We had a squadron member who had a brand-new truck. He really wanted that truck to be shipped. We were all caught up in things so that truck sat there for a couple of weeks until the battery blew,* so the truck wasn't functioning properly. I went with the squadron commander to this guy's house the day we moved his things, and after the packing we were going to see about getting his truck shipped. So, [this pickup] truck was literally full of a couple feet of ash. We had to dig it out. It was on a street that had been plowed by the snow plows. So, we had to dig it out from where the ash had been piled up around the sides of it. Spent about an hour just digging the truck out, and then it wouldn't start. So, we had the government vehicle, the truck from Cope Thunder, and we were trying to do [jumper] cables and everything. It would start up and then it wouldn't work, so finally what we decided to do is push the thing. So, here we had this guy's brand-new truck and because we are on the back roads of the housing area at Clark it was like being in the mid-west in a snow storm, in terms of the conditions, it was slippery. The cars would just slide around. So, we decided we would try to turn this thing around and just push it back to civilization where we could get a new battery or just ship it with the battery that didn't work. It was the squadron commander and me trying to get this guy's truck out, and we worked for like three hours and finally we got it dug out and turned around, and my job was to push this thing. I said I didn't want to drive the truck and he said he would take responsibility for driving the truck to be pushed and you be the pusher. OK, so it took us about half an hour to push this thing, just imagine in the ice or snowy weather where it is slippery. I'm taking this giant government truck and literally banging or ramming it into the pickup and then it would coast and I'd have to come up behind him and ram again. I'd try as hard as I could to just come up slowly to him and just push it, but in those conditions. Just things like that that you were doing in those conditions were so surreal. You'd never do that in the real world, you'd leave it and call a tow truck to tow it. It was survival instincts but we also wanted to help people and we had gotten to the point where we just had to do the best we could under the conditions. And

* *A common occurrence when ash blew under the hood and piled up on the battery. Eventually, the battery would short out and be ruined.*

so his conditions were that he got a truck that was a little bit banged up in the back, but he got his truck back."[26]

Evacuated Ash Warriors were given a toll-free number to call to tell the Air Force what they wanted done with POVs they left at Clark. Essentially, three options were available. They could abandon the vehicle and claim its value to an insurance company or the government, or they could have the vehicle shipped to them, or they could have someone at Clark sell it for them. The last option required approval of the government of the Philippines. On July 19, they approved a US request that each service member be allowed to sell one vehicle to local nationals. The agreement was a good deal for all parties. The Ash Warriors were able to sell a car without making an insurance claim, which was good fortune for those who had insured their vehicles with local companies. Many local insurance companies closed their doors after the eruption and never reopened. Local Filipinos normally could not buy American cars from the service people, so the cars sold very quickly to an eager market.

At Subic Bay and Cubi Point, the two Navy bases were clogged with abandoned cars as well. At Clark, unwanted or ruined vehicles were not shipped so that money and time would not be spent on low-value POVs. However, at the Navy bases, the policy was to ship them all, no matter the condition, to clear the roads and parking areas so clean-up could commence.[27] An evacuee at the Navy base, Air Force Lt Col Skip Vanorne, was selected to dispose of the vehicles. His qualification to do the job was simple. He wrote, "I got picked because I was still there. [Maj] Rob Parker and I were it, with only a couple of other folks still there.

"Now, what did we have to start? First, as everybody was leaving, they checked out with their Shirt or rep.* They gave us power of attorney to ship their car, the keys, the car's location, and addresses and phone numbers. The only other thing we had was about three thousand cars scattered over two bases, covered in ash, that had been parked for two weeks or more. The Subic commander was getting upset about our vehicle littering, and wanted them moved. We needed to clear the decks of cars, and in two weeks the *American Condor*† was coming, and we were challenged with putting a thousand cars on it while it was docked for a week. We were off to the races.

"The planning started by first figuring where we could prep and park that many cars. Clearly, we had to have almost all of them ready by the time the ship arrived. The only place big enough was the airfield at Cubi, so we got permission to use part of the taxiway and ramp space on the north side of the field. It wasn't much, but it had a fire hydrant that we could tap into for cleaning cars. We commandeered the vehicle ramps from the vehicle shipping/inspection station there at Subic, and had them moved to the taxiway. We added a fire hose for the hydrant, several tent shelters, power, vacuums, and lots of boxes and customs forms. Next, we used the contracting officer to start hiring drivers, cleaners, and boxers, and oh yeah, locksmiths.

"Our plan . . . was to spend the first week moving cars over to Cubi, and then the next two weeks (the one before the *Condor* arrived, and the one while it was there) cleaning/processing and loading, while the final effort continued, right up until the boat left. That last week, there were three separate actions taking place: finding cars and driving them to Cubi; cleaning and processing; and driving them back to Subic and loading them on the boat.

Every effort was aimed at one goal: pack out before the rains came, even if it meant working in the dark!

"So first we had to get cars over to the ramp [at Cubi Point]. Logistically, that simply meant taking a key from one of a thousand envelopes (note that that number is much less than the cars we had), finding the car somewhere on Subic, or maybe even Cubi (hoping the last known position was still correct), cleaning off enough ash to drive it, and move it to the ramp area. We had gotten the Clark SPs‡ to send us a complete listing of vehicles registered at Clark, with owners names, and license plate numbers. That helped a lot when it came time to ship that car, and the owner had failed to leave us any information – or keys. We quickly decided to take two teams, each with half the

* Slang for a squadron First Sergeant, a "first shirt."

† A ship designed to carry vehicles.

‡ Security Police.

envelopes, and a copy of the SP vehicle listings, and started driving around the base. We'd pick a vehicle, check the listing, [and] look for [its] envelope. If we found a match, one of our hired drivers in the back of the pickup truck would drive it over to Cubi. We kept a shuttle going that way while we picked the low hanging fruit. It didn't take long before we got down to no keys and a car, or a car and no keys. And in some cases, no car or keys. A couple of folks got a start on a second career [car theft] that week. We hired a couple of locksmiths, and trained two individuals in skills that they could use if they left the military to [become] car thieves. They got quite good at breaking into vehicles using slim jims. Then the locksmith would actually make a key for the car for us, and we'd drive it over to Cubi. Did I say drive? Many times it took a jump start or other minor maintenance to get them running. Remember, by this time some of these vehicles hadn't been started in three weeks, and the last I dealt with would be another three weeks, six in all.

At first it was take everything I could find keys for, then it was those I could make keys for, then it was almost what I could push

"And what happened when we had a key, and specific directions to where they had last parked the car, and it wasn't there? It turns out that the good Navy Captain Commanding Officer was getting more and more upset with our Air Force remote parking lot, that he started having some vehicles towed to clear away certain areas. And they didn't tell us where they put them. Lots of fun!

"The operation at Cubi was shaped like an upside down U. First stop was the hydrant and ramps. All the remaining ash was hosed off. No small garden hose for us, it was a manly operation. The three-inch fire hose made short work of any ash still on the car. It also made short work of a couple of paint jobs, too. A couple small paint peels got a lot larger when that water hit them. We were initially told that there could be no ash, dirt, debris, or other animal or vegetable products on the car prior to shipment. We reminded the inspector that the ash had been heated to something in the area of molten rock at one point, and we didn't think there would be any living anything still on the car after [going] through all of that. He finally agreed when he saw that given all the ash blowing around, the cars would never stay clean anyway, and we were actually doing a pretty good job of removing all of the ash anyway.

"A short drive up the hill began the emptying process. When you ship a car, you can't ship anything in it, so all personal belongings had to be removed, labeled, boxed, and prepared for shipment, themselves. We found everything that was left behind. Many folks had taken their prized possessions, expecting to return to Clark in a week or two. Much of what we boxed up was left-over clothes, car items, blankets, and the like. We found handguns and rifles, stereos, you name it. The most memorable, was a big family van. In the back was a large, cedar, hope chest, filled with wedding pictures, diplomas, family albums, and more. This one stood out because it was a USAF Academy grad,[*] and it had in it his football pictures, taken with the Air Force coach, and his saber.[†] The chest was literally packed with all those irreplaceable items from their life, and they had to leave it behind, with no knowledge of whether they would ever see it again. As such, we made a special effort to make sure theirs and everyone else's stuff was carefully packed, and that it all made it from the car to a sealed box.

"Once the car was empty, it was moved to the cleaning station, where the inside was cleaned of all dirt/ash/trash. Starting back down the hill, was the processing station. With whatever information we had, we filled out the forms, and processed it, inspecting damage and all. Once finished there, the car was driven down the hill, to the bottom of the taxiway, and parked ready to ship.

Lahars that crashed across the base exchange parking lot moved entire rows of cars into each other.

* As was Vanorne.

† Officer's sword. A treasured symbol of an officer's commission.

"Simple, right. All done in July, on a ramp covered in ash, blowing most of the time. It actually was a pretty smooth operation. When it was all over, we put just over 1,000 cars on the boat. We pushed a couple, but they got on the boat, and off Subic.

"All the cars went to a West Coast receiving location. The plan was that by the time of arrival, the PCS* orders for owners would have been worked, and the manifest lists were matched to the orders, and then the cars moved on. And if I remember right, we actually put one of our folks on the *Condor* to hand-carry the manifest to the States.

"As far as the cars I left behind, the transportation officer that finally came our way took over [so that I could return to Clark to] close the 1st Test Squadron.† They moved the ramps back to the real base inspection/shipment area, and worked what was left behind. More orderly I was told, but probably not near as much fun.

At the end of the day the AAFES cash register was $700,000 fatter, and many Ash Warriors were smiling

"How did I know which ones to ship? I didn't, and actually didn't care. At first it was take everything I could find keys for, then it was those I could make keys for, then it was almost what I could push. We pushed a Pinto [onto the *Condor*] that should have been pushed overboard. It had the bottom of the radiator torn out of it, and would overheat in a short drive. And it had a spray can paint job that looked hideous. But it was somebody's car, maybe an airman's that had spent his last buck on it, and he had the same right to get his car back as the guy with the Mustang. There was no way for us to know if these were intended to be left behind, so since we had no instructions for the ones we moved, it was assumed that they wanted them. I seem to remember a couple that we were asked to dispose of or sell if we could, but those got left for someone else to deal with.

"Now a couple of more stories. [Filipino] drivers were easy to come by, if you didn't check to see if the ink was dry on their driver's licenses. When the call went out to hire them, they came in large numbers. All [who wanted the job] were excellent drivers, with much experience. That's what they told us, anyway. I think we only ruined one clutch, on an RX-7. It was slipping so badly when they told me about it, we almost had to push it to the boat from Cubi. The Filipino driver said he didn't know what to make of it!? And you should have seen them lining up to move and drive the Ford Mustang with the 5.0 liter engine. I couldn't stand the chance of destroying that car, so naturally I had to drive it. But the best story was the car that left the processing station to be parked at the bottom of the hill. Someone noticed that the car was barely moving, about a hundred yards from the processing station. I ran over to it, able to keep up with it at a slow walk. I tapped on the window (closed tight, and tinted as dark as night). The electric window slides down, and inside the stereo is loud enough to be heard at Clark, the A/C is going full blast, and the driver, hands at 10 and 2 on the wheel, is [coasting] down the hill, at all of 1/2 a knot, oblivious to the fact the engine has died. We had a long laugh about that one later that night.

"And what of the Filipino packers and workers? Did they try to steal anything? Mostly not, but a few tried. One of our customs inspectors was an SP, topping out at about 6'2" and 220. At the end of every day, he patted down all the workers before they got on the bus to be driven back to the Subic main gate. About the most we ever found was a cassette tape. It got to be a joke with [the Filipinos] after a while.

"And I almost fired the contracting officer that was working for us. Seems he decided to hire a secretary to help him with the Forms 9 and such . . . hired to keep up with some of his contracting stuff, next thing we knew she was living with him in the Q. It must be nice when you can get the AF to pay for a live-in! We saw it all, it seems.

"And one about my drive back up to Clark the first time. There was an area that was about a foot deep in water that we had to drive through to get back. Enterprising Filipinos . . . were lined up at the shallow end, ready to guide you through the waters like Moses at the Red Sea. They told you to shut down your engine, they plugged the tail pipe with a rag, and started pushing you through. You steered, or ruddered, as you went, with a lead dog walking the road so you wouldn't wind up in a ditch. Half way into the deep stuff, water was coming into every hole in the car, some like little geysers in the floor board. Once through the lake, they unplugged the tail pipe, cleaned out some of the water, took most of my money, I fired that Muther up, and they bid me bon voyage."[28]

** Permanent change of station.*

† Vanorne had been assigned to Clark from Kadena Air Base, Japan, only a month earlier to take command of the squadron. His household goods were delivered to his house at Clark two days before the evacuation. He and his wife never unpacked the goods; they only opened enough boxes to retrieve things they needed to evacuate.

Many officers at Clark carried car and household goods insurance with a company in the United States that catered to military officers. The company sent a team of adjustors to Clark who settled many claims on-the-spot. Of course, many vehicles, personal and government, were damaged beyond repair. Even those that were operable were badly damaged by the ash. It filled window channels and every other place the water took it. Those who drove their POVs in the ash soon discovered that the abrasive stuff quickly wore out brakes and clogged radiators as well as air and oil filters. It was impossible to clean it off painted surfaces without damaging the finish, and after a few weeks it was unusual to see a vehicle that did not have some rust damage.

FIRE SALES

The Army and Air Force Exchange Service (AAFES) decided that it was cheaper to sell everything in the base exchange rather than pack and move the entire store's inventory. Although the large store was slightly damaged, almost all the inventory in the store and its warehouse was intact. On July 20, the doors to the BX opened for the first time in a month, and 200 Ash Warriors entered the store where everything inside was on sale at 25 percent of the marked price. Strict rules were in effect. The first 200 to enter the store were all active duty members of the mission essential force; each was selected by a lottery held the previous day. The lottery was held so that shift workers, especially the security police, would have an equal opportunity to shop. As soon as the first 200 entered the doors, a very long line formed as others waited his or her turn. As each smiling shopper left, with shopping carts stacked high, another was allowed in. Each shopper was only allowed to buy one of each high-value item: stereo, television, VCR, etc. By the end of the day, when some of the senior officers entered the BX, the place looked like a swarm of locusts had inundated it. At the end of the day the AAFES cash register was $700,000 fatter, and many Ash Warriors were smiling.[29]

Personnel took some much needed rest and relaxation at the base pool.

At the same time, Lt Col Doug Sobin, the morale, welfare and recreation (MWR) commander, who knew a good deal when he saw it, arranged a similar sale. He and his people cleared a large area on the first floor of the 3d Combat Support Group headquarters building then filled it with recreation gear from Clark, Camp John Hay, and Wallace Air Station. For the next six weeks they sold off the recreation supplies: golf clubs and balls, scuba gear, saddles, sports clothing, fishing poles, bowling balls, etc. For the first two weeks everything in the makeshift store was 25 percent off. The second two weeks everything was 50 percent off, and the last two weeks it was 75 percent off. Sobin reasoned that it would be more cost effective to sell the goods, even at reduced prices, than to attempt to redistribute the items to other Air Force bases. To redistribute the goods, he would have had to determine where the items were needed, pay to have the items packaged and shipped and risk loss and damage in transit. The store quickly became a popular stop for tired, dirty Ash Warriors coming off duty. The sale became a source of entertainment as troops teased each other with comments such as, "You'd better buy it this week, or it'll be gone before it goes down another 25 percent!" The sale took in $381,000.[30]

WHERE IS EVERYBODY?

Lt Col Mike Jordan had taken command of the 3d Mission Support Squadron (MSSQ) from Jim White in the Dau alternate command post amidst Pinatubo's eruptions. White, who had ended his tour, was returning to the United States, and Jordan recalled, "it was quick and dirty." A primary function of the MSSQ was to manage the tracking of all military and civilian employees assigned to Clark. When Jordan took command, most people were in the chaos at Subic Bay where the first sergeants and squadron commanders labored to find out where their troops and families were sheltered. For the next four months, the task of maintaining an accurate count, by name, of everyone's location was an enormous task. The main personnel tracking system was maintained on a computer, the Advanced Personnel Data System (PDS), but the PDS suffered the same fate as most computer systems on base. It was ruined by a combination of water, mud/ash, mildew, and tropical heat. Jordan and his troops had not "the time or the talent to fix it." As an alternative, they built a data base on a personal computer using a roster which had been produced the day of the evacuation, June 10. They then sent each unit its section of the roster, and the unit commanders would reply with any changes in manning. The ultimate "head count" became the only way to track people. All other personnel management actions were dictated by this data base which became known as the Mission Essential List. It became widely known as their "Bible" and without it they surely would have failed.[31] As the base gradually closed, fewer Ash Warriors were needed, so they were given assignments to other bases. A team of assignment experts traveled to Clark from the AF Military Personnel Center, the PACAF assignments office, and MAC. They provided assignments on-the-spot with orders* to go with them. Some orders were handwritten, but nearly all Ash Warriors received their assignment of choice. Mikkelson, the mounted horse patrolman, recalled being amazed that it only took less than a minute for him to be given his first choice.[32]

FUN

Few things are more important than mail to a service member who is serving overseas. Clark's postal workers, Detachment 3, 6005th Air Postal Squadron, commanded by Maj Rick S. Huhn, had evacuated to Subic Bay but were called back to Clark as soon as some of the dust settled. They used the MSSQ's mission essential list and set up new mail boxes for everyone in a small post office that was relatively undamaged. However, by the time mail started coming in, they were immediately overwhelmed with the backlog that had been building. By the time the postmen received all the mail that had been waiting to get to Clark they were faced with an avalanche of 300,000 pounds.[33] They put out a call for volunteers, and dozens of Ash Warriors trooped to the sorting room during off-duty hours to sort mail. They cross-checked every piece of mail against the mission essential list. If the piece was addressed to someone still at Clark, it went to the person's post office box. If not it went back to the United States. Outgoing mail became very important, and usually long lines snaked through the post office lobby as people waited to mail box after box of personal items back to the States. Although there were 15,000 fewer people at Clark, the outgoing mail volume was tripled. There were also 70,000 pounds of Air Force, official mail to sort as well. Maj Candy Shaefer, chief of information management, and her first sergeant, Danny Somers, arranged a unique system to get light into their powerless building so they could sort the backlog. They rigged mirrors in a way that reflected outside light through the windows so they could get the job done.[34]

Everyone looked for ways to get out of the ash. Base leaders organized three-day recreational tours to Wallace Air Station which had suffered very little ash fall. Wallace was clean, green, and had access to beaches on the South China Sea. Colonel Grime, the wing commander, directed that no officer could take one of these tours until every enlisted member who wanted to go had a chance to go. During the month of July, nearly half of the mission essential team took advantage of these tours. Transportation to and from the site, lodging, and meals were free.[34]

After MSgt Judy Sanders restored the Airmen's club, she went to work on a restaurant at the former golf course dining room. The place, named Jose's, opened in July. It was open to all ranks for lunch, and then became an officers' club after 5:00 p.m. Jose's became a usual place for Ash Warrior officers to gather for dinner, a cold San Miguel beer, and a game of crud.†

Some accused the Ash Warrior force of having more fun than they were actually having. A major US newspaper ran an article criticizing the base for using earthmoving equipment to clear the golf course while the runways remained buried in ash. However, the earthmovers that were on the golf course were clearing ash and mud from the drainage that ran through the middle of the course. It was the same drainage that had turned into a rampaging river that slammed through the commissary area on Black Saturday. The golf course was never cleared although some of the former caddies shoveled off a few of the greens in an effort to entice business back to the course.[36] A few diehards, General Studer among them, played the "course" in combat boots and shorts, hitting their shots from squares of carpet laid in the ash.

* *Assignment orders are vital. Without them an airman can accomplish none of the tasks attendant to a move: claim household goods, move his family, ship or claim his car, etc.*

† *Crud is played on a pool table. Only two balls are used, a cue ball and a colored object ball. A player rolls the cue ball by hand with the goal of either sinking the object ball or keeping the object ball in motion. Once he has rolled the cue ball and it contacts the object ball, his opponent then picks up the cue ball and takes his turn at either sinking the object ball or, failing that, hitting it so that it stays in motion. If a player fails to keep the object ball in motion he loses a "life." When a player loses three "lifes," he is eliminated from the match. The game is typically played by two teams of any number players. When one team eliminates the other team, the game is over. Losers must buy the winners a beverage of choice. It is understatement to say the games are spirited. Minor injuries are not uncommon as players must run constantly around the table to retrieve the cue ball.*

PASSING GAS

Over four million gallons of jet and vehicle fuel were on the base, and most of it was stored in a large "tank farm" on the southwest side of the base. The effort presented some unique challenges. While the inventory of fuels had been drawn down as part of an earlier conversion from JP-4 to JP-8,* over three million gallons of jet fuel were on hand. Another million plus gallons of motor gasoline (MOGAS) and diesel fuel were also at Clark. The large storage tanks and on-base fuel lines survived the eruption and many earthquakes. However, the pipeline between Clark and Subic was washed away at two river crossings. Lahars had washed through the line. The initial plan was to sell the fuel directly from the tanks to the Philippine government.[37]

However, by the first of October, Mount Pinatubo was still spewing ash and lahars were slamming down the watersheds. The ash was hundreds of feet thick near the mountain, and it was obvious that lahars might reach the tanks before it could be sold and moved.† If that happened, the tanks would surely rupture and the million gallons of jet fuel, MOGAS, and diesel might flow right through the middle of Clark and then into Angeles City. Some suggested that the fuel should be burned in the tanks. However, fuels experts quickly snuffed out that idea. They predicted that, even if those selected to ignite the fuel itself survived, the base and Angeles might be inundated with burning fuel.[38]

The logical choice was to truck the fuel to Subic Bay where it could be turned over to the Navy. Clark's fuels people, who had done yeomen's work keeping generators fueled during the frustrating efforts to re-establish water pressure, found and hired local contractors. Soon, fourteen 10,000-, 8,000-, and 5,000-gallon trucks were hauling fuel to Subic. With most of the river bridges out, this was a very difficult and dangerous task. By mid October, however, over two million gallons of jet fuel and MOGAS were transferred. They would have moved more, but some of the tank-bottom fuel was inaccessible because mud flows had already covered the tank-bottom drains.[39]

FINAL TALLY

At its peak, the Ash Warriors, Clark's mission essential force, numbered some 2,400 people. Nearly half were security forces charged with protecting American lives and property. The remainder devoted themselves to saving the personal possessions of their comrades and the military equipment of their nation. The magnificent effort of these selfless men and women, serving in nearly impossible conditions, cannot be overstated. On June 16, 1991 a small band of them returned to the Air Force's largest installation outside the continental United States; 164 days later they left after saving the following.

- The lives of over 17,000 American men, women, and children
- Personal possessions in 3,099 homes and 1,903 dormitory rooms
- 273 truck loads of munitions weighing 3,250 tons
- Over 2 million gallons of jet fuel, diesel and gasoline
- Nearly 1,500 40-foot tractor-trailer loads of equipment

** A world-wide USAF conversion. JP-8 is less volatile than JP-4; therefore, it is safer to handle and less dangerous in crashes.*

† A lahar, estimated at 25 feet high and 200 feet wide, swept through the southwest edge of the base the first week in September. (Source: 3d TFW SITREP, 090945Z Sep 91)

NOTES TO CHAPTER 8

1 Interview, TSgt Renee L. Longstreet with author, January 8, 1999, audio tape at HQ USAF/HO, Bolling AFB, DC [hereafter cited as Longstreet interview].

2 Longstreet interview.

3 Philippine Flyer, Magmatic Daily New Tremblor [sic], No 5, July 30, 1991.

4 Philippine Flyer, The Last Days Chronicle No. 8, August 2, 1991, p 1.

5 Ibid., The Last Days Chronicle No. 8, August 2, 1991, p 1.

6 Longstreet interview; Interview, Capt. Erin C. Gannon with author, January 6, 1999, audio tape at HQ USAF/HO, Bolling AFB, DC [hereafter cited as Gannon interview].

7 Longstreet interview.

8 Philippine Flyer, Magmatic Daily News Tremblor [sic], No. 14, July 16, 1991, "Dustbusters of the Day," p 1.

9 Gannon interview.

10 Grime interview 1992.

11 Longstreet interview.

12 Ibid.

13 Casey interview.

14 Message, 3 TFW SITREP, 290900Z Jul 91.

15 Grime interview 1992.

16 Clark AB Closure Report, p 37 (in Appendix 1).

17 Faulhaber, Col. Kenneth B., Air Force Journal of Logistics, "Clark Air Base Versus Mount Pinatubo," Fall 1992, p 12 [hereafter cited as Faulhaber].

18 Letter, Col. Alvin C. Schweizer, II to author, June 25, 1995.

19 Message, 3 TFW SITREP, 220900Z Jul 91.

20 Message, 3 TFW SITREP, 020945Z Sep 91; message, 3 TFW SITREP, 230845Z Sep 91.

21 Clark AB Closure Report, Hospital, November 26, 1991, p 27.

22 Philippine Flyer, The Last Days Chronicle No. 7, August 1, 1991, p 1.

23 Faulhaber, p 12.

24 Letter, Col Steven R. Miller to author, July 10, 1996.

25 Briefing, 3d Supply Squadron to 3 TFW commander, November 1, 1991.

26 Gannon interview.

27 Clark AB Closure Report, p 37.

28 Letter, Lt Col Ronald W. Vanorne to author, January 20, 1999.

29 Philippine Flyer, Magmatic Daily New Tremblor [sic], No. 17, July 22, 1991, "How about that sale?" p 1.

30 Clark AB Closure Report, p 32.

31 Ibid., p 31

32 Mikkelson interview.

33 Clark AB Closure Report, p 34.

34 Philippine Flyer, Magmatic Daily News Tremblor [sic] No. 4, July 3, 1991, p 1.

35 Briefing, 3 MWR squadron to 3 TFW commander, slide neither dated nor numbered, PACAF Historian's Office, Hickam AFB, Hawaii.

36 Grime interview 1992.

37 Faulhaber, p 12.

38 Ibid., p 13.

39 Ibid.

Chapter 9 TURN OUT THE LIGHTS

As mentioned earlier,* US negotiators and politicians were unwilling to pay what the government of the Philippines was asking for Clark and the Navy bases even before Pinatubo erupted. Once both suffered millions of dollars in damage, the United States was even more reluctant. Also, the volcanologists were giving no indications that the volcano would soon cease its activity. As long as the mountain continued to erupt, the danger of airborne ash to aviation was obvious. Even if the United States were to restore Clark to operational status the airspace was useless for training, and the ash, lahars, and pyroclastic flows had inundated the Cope Thunder ranges in Crow Valley.

Another factor that made an agreement seem unlikely was the opening of negotiations with the government of Singapore. In fact their government invited the United States to establish a military presence there. They offered buildings for Air Force squadrons, housing, and, most importantly, airspace for training flights. To the Navy they offered an alternative deep-water port to replace Subic Bay.

General Adams, the PACAF commander, recalled that the Philippines thought they had an undeniable position. In his word, they thought, "[if the Americans leave the Philippines] it will be the end of your presence in Southeast Asia, that's why you must stay here, and you must be willing to pay any price we demand." Clearly, the introduction of the Singapore solution, when added to the Pinatubo disaster, blew the Philippines' position out of the water. Adams said, "It was an interesting political fallout."[1]

Ambassador Richard Armitage and his team presented a final offer at the eighth and final visit of the US delegation in July. An agreement in principle, reflected in a joint statement issued on July 17, was reached. Its major provisions were as follows:

- The Philippines accepted a ten-year duration for operations at the Subic Bay Navy bases.
- The United States agreed to turn over Clark Air Base by September 1992.
- The Philippines accepted $203 million in security assistance monies along with other compensations that boosted the total to $362.8 million.
- A continuation of other non-defense initiatives.[2]

Following this announcement the United States sat back and watched as Filipino politicians argued in their Senate whether to ratify the negotiators' agreement. Some US senators formed a small coalition that saw no future for the United States in the Philippines, and they publicly ridiculed Philippine negotiator Manglapus for accepting less than he had boasted he was sure to get. However, the United States carefully avoided any official comments beyond that which they had agreed to in the July 17 announcement.

President Corazon C. Aquino's administration lobbied vigorously for ratification, but their efforts fell short. In September, the Philippines Senate rejected the agreement outright. The Senate "was a tough audience, consisting in large part of people who suffered under the

** Chapter 1 contains a discussion of the Philippine bases negotiations.*

US airmen raised the American flag over Clark AB for the last time.

rule of Ferdinand Marcos, a dictator seen by many in the Philippines as having been propped up by the United States solely for the sake of military bases."[3]

Even though the Senate rejected the agreement, few Filipinos doubted that the United States would actually abandon the bases under any circumstances. Citizens across the archipelago wanted the bases to stay, and Aquino threatened to hold a national referendum to set aside the Senate's rejection. As Armitage said, "In a very real sense, therefore, Filipino negotiators and senators probably felt quite secure in believing that somehow the United States would find a way to stay, to forgive their unruly behavior, to shower them with gifts, and, in general, to patch things up."[4]

...Filipino negotiators and senators probably felt quite secure in believing that somehow the United States would find a way to stay

It never happened. A joint announcement was made, after the Senate rejected the treaty, that all US bases would close by late 1992. Having seen the political handwriting on the wall, the 3 TFW staff had already set in motion a planning effort to be out of Clark as soon as possible. The first planning effort indicated that the base could be closed by March of 1992.

Some significant changes in leadership at Clark occurred as the Ash Warriors struggled in difficult circumstances. On July 6, 1991, Col John Murphy (Whopper),* who had planned behind the scenes for an evacuation, proudly stood on the ash-covered parade field and relinquished his command of the 3d Combat Support Group to Col Bill Dassler (Gladiator). Dassler, in turn, relinquished command of the 3d Security Police Group to his deputy Col Art Corwin (Pioneer). The ceremony, described in the *Philippine Flyer*, as the "Mother of all change of command ceremonies," featured fire trucks shooting streams of water into the air, and the mounted horse patrol carrying the American and command flags across the ash. A month later, on August 8, 1991, Col Jeffrey R. Grime (Eagle)† relinquished command of the 3d Tactical Fighter Wing, Clark's host unit, to Col Bruce M. Freeman (Hunter), who had been General Studer's Thirteenth Air Force vice commander and ramrod of the Joint Task Force-Fiery Vigil staff.

When the closure announcement was made on August 16, the Air Force had over a year to withdraw from Clark. Planners there estimated they could get everything out by March; however, Freeman and his officers had other thoughts. The enormous base had a generous supply of equipment that Freeman felt was not worth shipping: tables, chairs, desks, filing cabinets, beds, dressers, etc. Nearly 100,000 such items were on the base. If they did not ship such items, the withdrawal could be shortened by half, and those items comprised less than 15 percent of the value of the base inventory. To the Ash Warriors it seemed unreasonable to work an extra three months threatened by lahars and eating ash for so little return. Freeman started an effort to have the items declared as excess to US defense needs so they could be given to the Philippine Air Force, a declaration that could be made only by the US Congress. Staff officers at PACAF and the Pentagon, in what Freeman described as a "herculean effort," pushed the initiative through the Department of Defense and Congress.[5] Approval was granted in mid October, and the turnover of excess articles started immediately.[6] Clark planners, who had already planned for such a contingency, produced a modified plan that would close the base in early December. Equipment movement was progressing so well, however, the date was again moved forward to the Tuesday before Thanksgiving.[7] The Ash Warriors focused their efforts on a common goal – Thanksgiving dinner with their families. Home, food, clean linens and laughing children danced in their heads.

The 1947 Military Bases Agreement between the United States and the government of the Philippines (GOP) stated that the GOP owned all permanent structures and those structures would revert to the GOP [if the United States ever left the bases]. The agreement also qualified the definitions of non-removable and removable property. Finally, the agreement gave the GOP the right of first refusal for all removable property declared excess to US defense needs.[8]

The five objectives of the drawdown were to: provide an orderly, legal and dignified withdrawal; detail essential time phased actions required to terminate USAF activities; provide for the turnover of assets on the non-removable property listing; identify and remove all non-excess removable property; and transfer excess removable property to the GOP. The critical factor in developing the drawdown plan centered

** Murphy went on to Hickam Air Force Base, Hawaii, to command the Army and Air Force Exchange Service in the Pacific.*

† Grime went on to become the Inspector General of PACAF and was promoted to Brigadier General the following year.

around minimizing the time involved. This was due to the risk to personnel and equipment from continued encroachment of volcanic hazards on the Clark community and the potential for security and labor relations problems as the US withdrawal progressed. Likewise, maintenance of basic services such as water and power was becoming increasingly difficult to maintain.[9]

To meet these objectives and critical planning factors, the closure plan was based on a 120-day drawdown with a start date of August 1, 1991. All units would phase out in less than 120 days and all facilities and excess equipment would be turned over to the GOP within the same period.[10]

The base was sectored into six general geographic "packages" so that one package turned over to the GOP every two weeks starting mid September 1991. All facilities in a given package would transfer on that date except those critical to continued US conduct of operations during the withdrawal.[11]

The Ash Warriors focused their efforts on a common goal – Thanksgiving dinner with their families

The Philippine Air Force/CABCOM (Clark Air Base Command) was responsible for security within their packages after turnover. USAF/DOD security personnel provided point security only on any mission essential facilities remaining within a Philippine Air Force (PAF) controlled area.[12]

When the turnover plan was approved, the Thirteenth Air Force Facilities Drawdown Programming Plan (PPlan) was developed. The PPlan's time-phased actions were developed to coincide with the turnover of the various packages. Organizations cut back services provided, removed all non-excess equipment items and reduced manning requirements as each package turned over. Also, organizations consolidated remaining services and personnel into the minimum number of facilities possible. The planned result was to have the majority of services terminated and all non-excess equipment removed by mid November.

In determining the status of equipment the following measures of merit were considered. Non-excess equipment had to be critical to DOD needs, accountable, serviceable and had to be economically viable to remove. Computers, weapon system spares, munitions, most communications equipment, special purpose vehicles, high value medical equipment, weapons, and mobility/chemical warfare defense gear met the non-excess criteria and were shipped from Clark Air Base. Excess equipment remaining in place for transfer to GOP included most office furniture, some powered/non-powered ground support equipment, general purpose vehicles and general support supplies. All base agencies made inventories of what they determined to be excess equipment. These excess defense articles were consolidated and submitted through PACAF for the congressional notification process. The excess defense article list met the congressional notification period, and the USAF was authorized to transfer these articles to the GOP free of charge.[13]

The 3 TFW transmitted its last situation report to PACAF on November 18, 1991. It stated that 99 percent of government property had been shipped and only 20 truckloads remained. The mission essential force, once at a high of 2,400 was at 517. The previous week had seen the largest single reduction in Clark personnel since the evacuation, including the last of the USGS volcanologists. The Philippine Air Force controlled 90 percent of Clark's land area.[14]

The PAF struggled to maintain control of the base. As US security forces withdrew sector by sector, looters flooded into the areas. The PAF responded with gunfire. During the last week of October, the PAF encountered an average of 15 intruders per day, and in nine instances fired a total of 95 warning shots.[15] Colonel Freeman was deeply concerned that the Ash Warriors would not get off the base before someone was injured by a stray bullet. Looters in Hill housing, the first area turned over to the PAF, stripped the houses of doors, windows, appliances, air conditioners, electrical wiring, fixtures, toilets, sinks, and plumbing. Base engineers had to work hard to maintain water pressure because looters ripped faucets and sinks out of the houses without regard to the gushing result.[16]

The volcano's aftermath continued to threaten the Ash Warriors. During periods of heavy rain, a lahar alert was passed across the base on the brick network. Everyone was to stay on high ground. One young airman, whose curiosity overcame his discipline to follow orders, went too near a lahar. Suddenly it expanded and the young man was swept up in the roiling muck of swiftly flowing ash-mud. He desperately grabbed a piece of metal fence at the last second where he held on until his buddies rescued him. Another security policeman, a member of the mounted horse patrol, recalled, "One night when we [were] out on patrol, we went past where we were supposed to, past the bank of the river, and actually walked out on the ash. The guy who is riding with me, said 'Get out of there!' because the ash started breaking. I turned around to get my horse out, [but] the horse had fallen into the layer of the ash. After my horse fell through, I was holding his head on my shoulder. And we are both keeping our noses up, so we wouldn't drown. [My partner] ran over to get help, because the fire department is already in the area. We [got] ropes and stuff out, and I tied them around the horse. Well, as we dug the mud away from the horse, more ash was just coming into the hole. It set a suction so we can't pull the horse out. We couldn't dig the horse out, so we dug a trench to keep the water away from the hole. The ash was harder than concrete. We had to use pick axes around the horse's legs to chip the ash away. It took four and a half hours to get the horse out. Towards the end, when just the horse's hooves were in the ash it was still like the horse was cemented into place. The name of the horse was Butterfly. The only thing we were hoping for was that it wouldn't start raining, because a flash flood would be the end of the horse."[17]

As Thanksgiving approached, last minute things were accomplished. November 20 was the last day to turn in equipment, and the next day was the last an American could sell a POV to a Filipino. On Friday, November 22 all the remaining Ash Warriors moved into Chambers Hall, the last remaining American billeting location. On November 23, the last mail to be postmarked with Clark's military zip code (APO) departed the base. Jose's, the small golf course restaurant that had served as the officers'club, saw its last crud match on the 24th.[18]

On November 25, dignitaries started to arrive. At 1600, General Jimmie V. Adams, PACAF Commander; his wife, Ouida; Col Pedro Rivera; Col Vincent Majkowski; Col Olan Waldrop; Maj Scotty Rogers; Capt Roy Joy; CMSgt Gary Kushner; and MSgt Rita Dzurenda arrived at Naval Air Station Cubi Point, where they were met by General Studer and Rear Admiral Thomas Mercer, the senior naval officer in the Philippines. Cactus, Clark's UH-1 helicopters, flew the party to the base's south helicopter ramp where they were met by Colonel Freeman.[19]

Visiting dignitary at flag ceremony.

The next morning, November 26, 1991, the American flag was raised over the Clark parade field where it had flown for nearly 93 years, interrupted only by the Japanese occupation during World War II. Over the tens of decades the base had seen every kind of military experience from cavalry to Phantom jets. Countless thousands of Army and Air Force service members fondly recall the barns, the beautiful parade field, magnificent acacia trees, smiling Filipinos, breath-taking monsoon rains and lifelong friendships. The last generation there has those memories as well, but they most remember of Clark uncertainty, fear, eruptions, earthquakes, evacuations, lahars, and ash. Nonetheless, they would have fond memories, too. The Ash Warrior spirit bonded friendships that will never die.

At 7:45 a.m., Frank Wisner, US Ambassador to the Philippines, arrived at the south helicopter ramp with his wife, son, and Mr. and Mrs. Hubbard. They joined General and Mrs. Adams and other Ash Warriors at the Airmen's Club for a brief reception. At this reception, Lt Col Kevin Collins was promoted to colonel in a brief ceremony. Collins was "pinned" with the silver eagles which had been given to his father nearly fifty years earlier by the people of Baguio. At 8:25 the official party went to the parade field for the official closure ceremony.

The ceremony started at 8:30. Among the dignitaries on the reviewing stand were Ambassador Wisner; General Adams; the Philippines Secretary of National Defense Davia; General Studer; Brigadier General Acot, the Clark Air Base Command commander; and Colonel Freeman. Across from the reviewing stand were two honor guard formations which represented the USAF and the PAF. Rain and some cleaning had removed most of the ash from the parade field, and the immaculate airmen stood in spit-shined boots and freshly laundered BDUs.

Awards were presented to General Studer, Colonel Freeman, Colonel Rand, and others. Nearly 100 members of the press covered the ceremony and crowded around the presentations taking pictures rapid fire.*

After the personal awards were presented, the honor guard formations marched into a cor-

* *Every Ash Warrior received the Joint Meritorious Unit Award.*

don in front of the reviewing stand. While Rand read a brief history of the 3d Tactical Fighter Wing, three riders from the mounted horse patrol rode at a slow, deliberate pace through the cordon with the wing flag as the formation saluted. When the riders reached the center of the reviewing stand, they wheeled the horses to present the colors to Freeman. No words were spoken. Freeman saluted his colors, and the horses, resplendent in shined tack and Air Force-blue foreleg wrappings, wheeled again and withdrew the flying colors from the parade field.* Rand then read a brief history of Thirteenth Air Force, and the process was repeated with the Thirteenth flag. General Studer saluted his colors that were presented by three other mounted horse patrol riders. It was the last official act for the Air Force's last mounted horse patrol.†

The final sequence was to take down the American Flag. An Ash Warrior honor guard slowly lowered the Stars and Stripes and folded it into the customary triangle. They then passed it to SrA Jeffrey Oakly of Baltimore, Maryland. Oakly cradled the flag in his arms and held the blue field of white stars against his chest as he marched slowly to where Wisner, Adams, and Studer stood in front of the reviewing stand. The flag was passed from Oakly to Wisner to Adams and, finally, to Studer. Studer moved a few steps to the side of the reviewing stand, an act symbolic of removing the colors from the battlefield, and gave them to SSgt Scott Clark who carried them from the parade field.‡ The dignitaries left the reviewing stand and the last Ash Warriors prepared to leave Clark.[20]

Five buses carried the last 170 Ash Warriors to Subic Bay where they loaded onto an Air Force transport, what they called a "Freedom Bird," and departed the Philippines for Andersen Air Force Base, Guam, the new home of Thirteenth Air Force. Colonel Freeman boarded a helicopter a minute ahead of General Studer, and the aircraft departed Clark Air Base at 1:22 p.m., November 26, 1991.[21]

Within a week of their arrival at Andersen Air Force Base, the Thirteenth Air Force staff was greeted by a devastating typhoon that swept across Guam.

** The 3 TFW was reassigned to Eleventh Air Force and moved to Elmendorf Air Force Base, Alaska. Usually, at a ceremony such as this one, the colors are furled around the staff and put into a long cover. Freeman would have none of that; he did not want "anyone rolling up his colors and sticking them in a bag." (Source: Freeman interview)*

† At the end of the ceremony, the riders removed everything from the horses and loaded the faithful animals onto trailers for transportation to Filipino owners who had purchased them at auction.

‡ The flag is now on display at the 36th Air Base Wing headquarters, Andersen Air Force Base, Guam.

NOTES TO CHAPTER 9

1 Adams interview.

2 Armitage interview.

3 Ibid.

4 Ibid.

5 Freeman interview.

6 Message, 3 TFW SITREP, 210300Z Oct 91.

7 Freeman interview.

8 Clark AB Closure Plan, p 17.

9 Ibid.

10 Ibid.

11 Ibid.

12 Ibid., p 18.

14 Ibid.

15 Message, 3 TFW SITREP 181115Z Nov 91.

16 Message, 3 TFW SITREP, 280530Z Nov 91.

17 Freeman interview.

18 Mikkelson interview.

19 Calendar, Clark Air Base Withdrawal/Closure, November 5, 1991, PACAF/HO, Hickam AFB, Hawaii.

20 Clark Air Base Final Departure Plan (CABFDP), Proposed Itinerary 6, November 21, 1991.

21 Description of the closure ceremony was extracted from the Freeman interview.

Chapter 10 A DISASTER IS A DISASTER

Perhaps it is human nature to deny impending doom and gloom. One says, "that hurricane, as the many before it, will miss my house." Then, in an effort to validate our nature to deny, we make no plans. Perhaps some are proponents of the "ignorance is bliss," school. Perhaps our hearts control our good sense.

Certainly, Mount Pinatubo seemed innocuous when viewed in comparison to Mount Arayat. Arayat, which majestically rises from the rice paddies 10 miles east of Clark Air Base, looked like a real volcano. It was the classic cone-shaped mountain whose appearance would lead even the most casual observer to the conclusion that it had once produced violent eruptions. However, Pinatubo to the west was nothing more than another big, green hill among other big, green hills. Scientists knew Pinatubo was a volcano, but there are many volcanoes along the Philippine archipelago similar to Pinatubo. Mostly they are quiet, and Pinatubo had not erupted for centuries. Former residents of Clark Air Base, who were interviewed for this book, universally proclaimed confusion when they first saw the steam which emitted from cracks in the far side of Pinatubo. Without exception, their reaction was "Volcano? What volcano?"

The Call for Assistance

Fortunately for the Clark residents, senior base officials, despite their skepticism, did not delay in securing the services of experts from the US Geological Survey. Just as fortunate, USGS scientists rushed to the scene and were evaluating the mountain only three weeks after the steam started. Bureaucracies are notoriously and ponderously slow to react, however, such was not the case at Clark and the USGS. They reacted quickly even though the mountain was at a relatively low level of activity when the call went out for help. There is no way to assess how much the disaster at Clark and Subic Bay might have been magnified if the arrival of the USGS team had been delayed by only a few weeks. The mountain erupted less than two months after they arrived.

Despite their success, the volcanologists are the first to admit that predicting an eruption is an inexact science. Nonetheless, they were spot-on with Pinatubo; 250,000 Filipinos, who lived close to the mountain, moved away from the mountain in the week before it blew. Fifteen thousand Americans left Clark 48 hours before the first "small" eruption and five days before the volcano destroyed itself in blasts that made it the second largest of the twentieth century.

General Adams, General Studer, and Colonel Grime praised the USGS-PHIVOLCS team as heroes of the effort in interviews after the US flag was taken down at Clark. General Adams said, "Thank God they were a lot more accurate than any of us thought they could be."[1] Colonel Grime praised the team by saying, "at the end, they are heroes in my mind. [I] couldn't have asked for guys with more guts, guys more willing to work the problem."[2]

Subic Bay also suffered heavy ashfall and typhoon rain.

Information Control

Former Clark residents may argue forever the questions of how much they knew about the volcano and how well the military bases at Clark and Subic Bay were prepared to evacuate. Senior officials at Clark were divided in their approach. Some wanted to make mass public service announcements parallel to the scientists' discoveries and openly plan for a mass evacuation. However, both the Thirteenth Air Force and 3 TFW commanders insisted that information be passed strictly through the chain of command and from a single source, the base public affairs officer. Further, they prohibited public release of information they viewed as incomplete or speculative in order to mitigate Clark's active rumor mill. The record is clear; many meetings were held with volcanologists and unit commanders, and the USGS team chief went on base television to discuss the dangers two weeks before the evacuation. Many interviewees do not recall reading about the volcano in the base newspaper or seeing the televised briefing. Several recall learning details about the volcano through their commanders and first sergeants.

Although evacuation plans were not published until three days before the evacuation, nearly 15,000 people moved off the base in an orderly manner, without injury, in about six hours. However, many did not take critical items such as marriage licenses, insurance policies, and prescription medications. During the days leading to the evacuation, there are no documented incidents, or even rumors, of widespread panic in the population. Many believe they would have been better prepared to evacuate if they had been included in the plans sooner. The Thirteenth Air Force commander stated that one of the main reasons he delayed evacuation was to give his people more time to adjust mentally to the concept of evacuating.[3]

Few recall it as being fun, but all remember the satisfaction of succeeding under grueling conditions

Nearly everyone interviewed expressed feelings of at least resentment, if not total anger, at not being told more sooner. They wrongly assume there was more to be told. Although some commanders wanted a more open approach to planning for a possible eruption, information was passed through the chain of command once the Thirteenth Air Force commander and wing commander were satisfied of its reliability. Of course, no one knew prior to the evacuation to Subic Bay that the eruptions would be so big or that a typhoon would strike simultaneously.

Without question, Subic Bay was unprepared for the onslaught of refugees which slammed against its gates on June 10. However, by June 12 everyone was sheltered and the initial chaos was under control. The Thirteenth Air Force commander attributes the chaos to Clark officials not giving enough precise information to Subic Bay before the onslaught and to "lack of leadership" from some senior Air Force leaders who were at Subic Bay.

Training and Discipline

Some observers of the evacuation, both military and civilian, have claimed that such an evacuation would not have succeeded but for the training and discipline of the military members and their families. Despite previous hardships and the cynicism and skepticism those hardships produced in Clark's people, they evacuated the base in good order very quickly. Most first sergeants and squadron commanders rallied and organized their people at Subic Bay despite the confusion and congestion, and they did it the old fashioned way – pencil, paper, sweat, and shoe leather. During the horrific conditions of Black Saturday, Navy and Air Force engineers left their families in homes and shelters to inspect buildings, clear roofs, and lead endangered evacuees to safer shelters. Good order and discipline saved many lives that "day" as buildings collapsed, in at least two instances, only minutes after evacuees were moved. Evacuees boarded ship after ship without incident or accident; and when tempers flared, officers and sergeants were on the scene to extinguish the sparks.

Navy Medics attended the evacuees.

One incredible aspect of the evacuation from Subic Bay to the United States is that no one was fatally injured and there were only a few minor injuries. More than 20,000 people climbed gangplanks; children moved up and down steep, steel ladders in the bowels of warships; elderly and infants donned flight gear to cross flight decks, which some claim is the world's most dangerous work area; thousands were transported on landing craft. Navy ships frequently practice transporting non-combatants, and that training proved its worth during the shuttle of evacuees from Subic Bay to Cebu. Their success is testament to the training Navy commanders gave their troops, the conscientious actions of those sailors, and the indomitable grit of the military families.

Many Ash Warriors described the challenges they endured as "one long ORI." Every Air Force base undergoes a periodic Operational Readiness Inspection* when headquarters inspectors evaluate every aspect of the base's ability to operate under combat conditions. Throughout the inspection, ORI team members inject problems with which the base must deal. Air attacks are simulated, and after each attack damage must be assessed and compensation made for the loss of facilities or capabilities. At Clark, the volcano tested the Ash Warriors for weeks on end: eruptions, rain, lahars, no water, no electricity, collapsing buildings. They reacted as they had been trained to react by assessing the damage, inventorying resources, and finding alternative ways to do the job. Few recall it as being fun, but all remember the satisfaction of succeeding under grueling conditions.

Communications

No commander can succeed without reliable communications. The brick system, a trunked land mobile radio net, gave Clark commanders instant communication with every Ash Warrior. The base possessed over 600 of the hand-held radios, or one for every two Ash Warriors during the eruptions, and one for every three or four during the base closure. When the CAT commander ordered a bugout over the radio and backed it up with the base siren, everyone moved immediately. Even when the base was without power, the brick system still operated from emergency generators. Nearly every commander interviewed for this book volunteered their belief that the reliable bricks saved many lives.

Planning

A plan to evacuate Clark was thrown together hastily with no time to practice or refine the details. The best to be said of it was that it worked. However, it is reasonable to assume that the evacuation would have gone more smoothly if base officials had started the effort sooner and more publicly. The base was not well trained in evacuation operations. Even though a plan existed to fly dependents out of Clark in a wartime scenario, the plan was never exercised on anything but a very small scale, so the families were not well rehearsed.

One of the most painful aspects of the evacuation discussed by every evacuee and Ash Warrior was the belief that they would only be gone from Clark for two or three days. Everyone interviewed, from the commanding general to the youngest airman, said they thought they would only be leaving the base for a very short time – until the danger was past or the mountain erupted and quieted down. However, no one recalls what information they got that would lead them to such a conclusion. The evacuation pamphlet suggests clothing for three days and pet food for three days, so perhaps evacuees latched onto that number and assumed the best. Certainly, many left things in their homes, pets and important papers, they would have taken had they believed they would be gone for weeks or would never return. The warning scheme

Willing helpers attend to a pregnant woman on her arrival at Kadena AB.

* *Names for inspections changed over the years. The Operational Readiness Inspection was later divided into two inspections, the Initial Response Readiness Inspection (IRRI) and the Combat Employment Readiness Inspection (CERI).*

used by USGS–PHIVOLCS gave no mention of an estimated time for the volcano to be "safe" following an eruption. The lesson here is clear. If an evacuation is ordered, evacuees should plan their "bugout" kit for the worst scenario – never return.

Some covert planning proved to be insightful. Essentially, the plan for the mission essential force (those who stayed at Clark) featured two fall-back positions. The first was the alternate command post at Dau, which featured air-filtered, redundant emergency generators, the main station for the brick system, air conditioning and large, open floor spaces. The second was the stockpile of food, water, and communications at the Pampanga Agricultural College. Each provided an additional margin of safety, more distance, from the volcano. It was a good scheme, well organized and should be a primary planning factor for similar scenarios.

Command and Control

Several levels of command were at play during the disaster. JTF-Fiery Vigil, commanded by General Studer, controlled all military actions in the Philippines. Studer's immediate commander was USCINCPAC in Hawaii, and Studer's immediate Air Force subordinate was Colonel Grime, the 3 TFW commander at Clark.

As events unfolded, Studer elected to command Fiery Vigil from Clark, the area he viewed as "the front lines," rather than from Manila or Subic Bay. Some officers on Studer's staff thought Studer should be more visible to all US military forces on the island, especially the Navy at Subic Bay, where thousands of refugees were sheltered.

Against this backdrop, two events happened that nearly cost Studer his job. First, Fiery Vigil staff officers arrived at Clark from US Pacific Command (USPACOM). Their job was to provide staff support to Studer at Clark, and one of their primary responsibilities was to keep USPACOM informed of events, requirements, and status of forces at both Subic Bay and Clark. When the officers arrived, they were put under the command of the 3 TFW CAT commander, Colonel Anderegg. However, Anderegg knew little of joint staff procedures and was totally immersed in conducting emergency actions in the face of continuing eruptions. Without strong leadership the Fiery Vigil staff was ineffective and produced virtually no reports to USPACOM. Second, during the bugout response to the first eruption, Studer's communication with the outside world was lost for a short time, an unacceptable situation in combat or disaster response.

USCINCPAC, understandably concerned, called Studer and asked him, "why I shouldn't fire you right now." Studer replied that the communication problem had been fixed, that he would fix the staff problem, and he still believed he should be at the point of the spear. USCINCPAC relented, and Studer remained commander of Fiery Vigil. Immediately after the painful conversation, Studer recalled his vice commander, Col Bruce Freeman, from Subic Bay and put him in charge of the Fiery Vigil staff at Clark. Freeman was very experienced in joint staff matters and quickly had them working efficiently while Anderegg again focused his attention on directing emergency response. During an interview in 1999, Studer reaffirmed his conviction that his place as commander, Thirteenth Air Force, was on the front line of the battle.

Senior commanders at both Clark and Subic Bay were effusive in their praise for the effectiveness of their subordinate squadron and group commanders. Almost without exception, those dedicated and brave professionals gave their very best. The security police group's reliance on decentralized control and unit levels of authority in an Air Base Ground Defense posture was superb according to every senior officer interviewed.

Caring for the Troops

Displaced Ash Warrior families scattered across the United States and settled in "safe haven" locations. The safe haven tag authorized payments and allowances to the families to carry them over until the Air Force could decide where to send them. As the families settled into safe haven locations they were totally confused as to what the entitlements authorized. Nearby Air Force bases tried to be helpful, but the Air Force personnel and legal systems were slow to respond to the evacuees' needs and sometimes put out conflicting guidance. In fact, entitlements changed four times after the evacuation. Each change was more beneficial to the evacuees, but the changes resulted in much confusion.[5]

Of course Air Force bases and families opened doors and hearts to evacuees, but many evacuees were not in safe haven locations near bases. Personnel and legal services immediately established toll-free hotlines to answer evacuees' questions, but everyone interviewed recalled weeks of chaos trying to find clothing and places to live followed by months of trying to track down the locations of household goods and vehicles shipped from the Philippines.

Summation

Many lessons were learned from the Pinatubo eruptions and subsequent closure of Clark Air Base, and most of these lessons are included in Appendix 1. However, words alone seem inadequate to express the suffering and trauma caused by Mount Pinatubo. American families were torn away from Filipino domestic workers that were thought of as part of the family. Teenagers left school on Friday and never saw the school again, and in many cases never saw their friends to say goodbye. Many family members later underwent therapy for post-traumatic stress disorder. The trauma that Black Saturday visited on the huddled masses of evacuees can never be measured. Many survivors weep as they describe the fear and frustration of the eruptions and evacuation. Some told of instances, years later, where they fled from tiered parking garages because they could feel the structures flexing, and the vibrations gave them "flashbacks" of the earthquakes on Black Saturday. Despite government funding for expenses and losses to household goods and vehicles, the disaster was a disaster, and, try as one might, there is no happy face to put on it. No amount of money or insurance can replace the displacement of friends and the loss of near-family. The trauma caused by nature's violence scarred everyone who was there with the knowledge that humanity is fragile on our fragile Earth. Violent natural forces kill many each year. Ash expelled from Pinatubo lowered temperatures in the northern hemisphere one degree centigrade for nearly two years.[6] When several Pinatubos happen at once, humankind will be challenged to survive. Clark survivors think of these things. They lost their innocence – on a Saturday in the Philippines.

NOTES TO CHAPTER 10

1 Adams interview.

2 Grime interview, 1999.

3 Studer interview.

4 Ibid.

5 Clark AB Closure Report.

6 Hoppe, Kathryn, "Mount Pinatubo cloud shades global climate," Science News, July 18, 1992, v 142, n 3, p 37 (1).

Afterword

CLARK, 1998

The author traveled to Clark in November 1998. He was amazed at what he found. President Fidel Ramos established the Clark Development Corporation (CDC) in 1993. The mission of the CDC is to encourage commercial development and tourism in a Special Economic Zone which encompasses all of the former Clark Air Base as well as a land mass as large as Clark to the north and west. The Special Economic Zone provides significant tax advantages to businesses that operate on the old base.

The central part of the base, that area north of the parade field and around the officers' club and Chambers Hall is now the Mimosa Resort. Mimosa Resort operates independently of the CDC. The resort encompasses the old golf course and club house, the junior NCO housing around the golf course, the officers' club, Chambers Hall, some of the barns in that area, and much of the new housing between the parade field and Wagner High School. The Mimosa Resort has been the focus of a $64 million investment of Taiwanese, Japanese, Korean, and Australian investors. The centerpiece of the resort is the former Chambers Hall, which is now a Holiday Inn. It claims to be a five-star hotel, and in the author's opinion it is close to that. A room there goes for $100 per night. The remodeling effort in Chambers is nothing short of magnificent. Next door, the former officers' club is now a casino. Several of the barns behind the casino are theme restaurants such as a Gasthaus and a Japanese restaurant. The junior NCO housing across the street from Chambers has been extensively remodeled into beautiful golf villas with tile roofs and arched Mediterranean doors and windows. The golf villas rent for $250 per night. A new club house has been built at the golf facility which now features two 18-hole courses, and a third is under construction. The old Clark course no longer exists; the holes were all redesigned to accommodate the expansion. Both of the new courses have been seeded with imported grasses, and the ash was used to mound and sculpt the fairways, tees and greens. The former caddies are gone as well – replaced by female caddies in the Asian tradition.

Col John Murphy's former headquarters building at the east end of the parade field is now the Clark Museum. The museum uses the entire first floor of the building to present a comprehensive history of the base from 1898 to 1998. The curator is Miss Cefey Yepez. The Fort Stotsenburg gate posts still adorn the center of the parade field.

Over 200 industries operate on the base. These are computer assembly factories and other high-tech businesses. Across the airfield, near where the Ash Warriors huddled in the Dau command post, is an enormous Yokohama Tire factory. There are also nearly 50 duty-free stores on the base. These are not small shops such as ones in airports, but they are large stores such as Nevada Discount Golf and others. All the duty-free stores are located in the area of the BX and commissary. There is a water-theme park for children and an amusement park in conjunction with Expo 2000. The exposition was built on the site of the former "elephant cage." Developers took down the antenna wires and built a large, free-form tent enclosure over the pilings from the elephant cage.

The old MAC terminal is now the terminal for Clark International Airport. The terminal has been given a $2 million renovation and is a first-rate facility. The airfield is fully functional and charter flights bring tourists and businessmen to the base. One amusing sight in the terminal area is the old mortuary which is now the Four Seasons Restaurant, complete with bars on the windows. The author ate lunch there twice and both times employees tried to convince him that the building

Thirteenth Air Force Headquarters building.

was never used as a mortuary, apparently they were losing business because of the building's former use. The food was great and comes highly recommended if one does not object to eating in a former mortuary. Tourism seems to be flourishing. Families come to Clark and stay either in the Holiday Inn or a golf villa. The men do business and play golf, while the families shop in the duty free stores and play in the amusement parks. Tour buses crowd the parking lots, and the Holiday Inn is very busy.

The only area of the base that does not look as good as or better than before the volcano is the Hill housing. It is closed off from the rest of the base. The area was stripped of doors, windows, electrical wires, plumbing and fixtures when it was turned over to the Philippine Air Force during the Americans withdrawal. Nothing has been done with it since.

Angeles City, Dau, and Mabalacat are bustling, thriving communities. Jeepneys and trikes crowd the streets, and it is hard to believe that nearby was an enormous base that once drove the region's economy. Many of the small, cottage industries are gone, but in their place are other businesses. Nearly 2,000 US military retirees still live in the surrounding communities. Jim Boyd still runs the Retirees Activity Office (RAO) as an unpaid volunteer. There, retirees can communicate via computer with DFAS and AFPC to work pay and records problems. The University of Angeles built a new hospital which takes CHAMPUS so retirees get medical care there. The RAO can be contacted at HYPERLINK mailto:rao_cabr@mozcom.com rao_cabr@mozcom.com (Note: there is an underline mark between rao and cabr.)

Another retiree activity that still thrives is the local chapter of the Veterans of Foreign Wars. Three years ago, the Clark cemetery was in disrepair. The cemetery is not an official overseas US cemetery, but it holds the remains of several hundred US and Philippine veterans. Although Thirteenth Air Force entered into an agreement with the Philippine Air Force to maintain the cemetery and provide flags for it, neither has fulfilled its part of the agreement for years. The VFW has taken over maintaining the cemetery. It costs them about $7,000 per year which they collect through fund raisers and donations. They decorate the cemetery each Veterans Day and Memorial Day. Donations and information requests can be sent to Mr. B.K. Hubbard, PSC 517, Box R-CV, FPO AP 96517-1000. Mr. Hubbard can also be contacted through email at bkhub@mail.ang.sequel.net

Appendix 1 CLARK AIR BASE CLOSURE REPORT

13AF PPlan 91-13-01

Facilities Drawdown Plan

26 Nov 91

Part IV: Commander's Comments

The programming plan provided good guidance. Adjustments were made as required in coordination with 13AF/XP and Clark AB OPRs. The support of all Clark agencies was superb. The drawdown plan, turnover procedures and documentation requirements were briefed and discussed with Brigadier General Acot, CAB-COMICC on 12 September 1991. The milestones for the Programming Plan and the inherent milestones associated with the phased turnover of sector packages drove the process through the withdrawal. We met the milestones for the sector turnovers and met the objectives. In terms of timeline tracking, we got a late start but caught up quickly. Overall, Clark agencies throughout assisted with the huge effort of packing and pick-up of supplies and equipment. Early phases focused on household goods, followed by government facilities equipment.

The Excess Defense Article initiative proved to be a difficult challenge. Congressional notification was required to transfer excess U.S. Government property to the Philippine Air Force. To accomplish this, an inventory was conducted of facilities, equipment we then listed with condition codes and forwarded through PACAFIXP to SAF/IARP for final packaging to go to Congress. Following a required 15-day congressional review period, approved property could be transferred. The size of the undertaking was almost unmanageable. People worked long and hard to develop as precise a list as possible. With limited numbers of people, damaged facilities, and just recording data, it was a project which took extensive time – more than we anticipated. An early start is essential. We obtained approval for the Excess Defense Articles to be transferred any time after 17 Oct 91, and the process proceeded on schedule.

The Base Closure Committee executed the plan and provided the "cross-tell" of information to address concerns, resolve disconnects in the individual organizational concepts and to propose coordinated and consolidated decisions and recommendations to the base leadership for base wide closure issues. It worked well, served as a clearinghouse, fixed problems early on, and focused attention – on time – on the "make or break" critical issues. We tracked milestones, calendar events, equipment turn-in schedules, vehicle utilization, personnel flows, Excess Defense Articles, sector package processing, security issues and the documentation to put it together. Throughout, the process and transition went just about as planned. The units across the base had good documentation to verify a record or event. Again, it was essential. Additionally, below are comments and some lessons learned from the major activities that executed the plan. The 120 days was a fast-paced operation that produced the desired smooth, orderly and dignified U.S. withdrawal from Clark Air Base.

GENERAL

On 17 July 1991, US and Government of Philippines (GOP) representatives publicly announced that the US facility at Clark AB, RP would be turned over to the GOP NLT 16 Sept 1992. Following this announcement, 13AF and 3TFW were tasked to develop a drawdown plan to remove the US presence from Clark AB. The 1947 Military Bases Agreement (as amended) stated that the GOP owned all permanent structures and these structures would revert to GOP. This agreement also qualified the definitions of non-removable and removable property. Finally, this agreement gave the GOP the right of first refusal for all removable property declared excess to defense needs. This agreement set the guidelines for drawdown objectives.

The objectives of the drawdown were:

1) Provide an orderly, legal and dignified withdrawal;
2) Detail essential time phased actions required to terminate USAF activities;
3) Provide for the turnover of assets on the non-removable property listing;

Air Force members packed property for removal from Clark AB.

4) Identify and remove all non-excess removable property;

5) Transfer excess removable property to the GOP.

The critical factor in developing the drawdown plan centered around minimizing the time involved. This was due to the risk to personnel and equipment from continued encroachment of volcanic hazards on the Clark community and the potential for security and labor relations problems as the US withdrawal progressed. Likewise, maintenance of basic services such as water and power was becoming increasingly difficult to maintain.

To meet the above objectives and the critical planning factors, the closure plan was based on a 120 day drawdown with a start date of 1 Aug 91. All units would phase out in less than 120 days and all facilities/excess equipment would be turned over the GOP within the same period.

The base was organized into six general geographic "packages" with one package turned over to the GOP every two weeks starting mid September 1991. All facilities in a given package would transfer on that date except those critical to continued US conduct of operations during the withdrawal. Description of Packages are as follows:

Package I - Mid Sept 91

Landmass east of runways
Munitions Storage Area (MSA) I
Golf Course
Hill Military Family Housing Area (MFH)
* 7500 Series dormitories
* In-flight kitchen
* CSG HO building

Note: Even prior to the eruption of Mt. Pinatubo, these facilities were prepared for turnover to the PAF starting in September 1991.

Package II - 1 Oct 91

Northern landmass of the base Hospital
6922 ESS and antenna farms
Bambam Drop Zone

Package III - Mid Oct 91

POL Storage Area
SATCOM Site
Sapangbato MFH
Various MWR facilities
MSA II

Note: Because the landmass of this package was most affected by mudflows/flood-mg, this package was designed for turnover to coincide with the onset of the "dry" season. Additionally, with this package, the PAF now controlled the entire perimeter of Clark AB.

Package IV - 1 Nov91

Runways
Flightline facilities

Package V - Mid Nov 91

MFES Complex
Commissary
Lily Hill MFH
Dining Hall Facilities
Supply Warehouses
Remaining Dormitories (except SP)

Package VI - End Nov 91

Parade Ground/Golf Course MFH
Various mission essential facilities no longer required
Mission Essential Facilities - Close Date
Billeting
Messing
SP/CE/Communication compounds.
Infrastructure

Note: These mission essential facilities were those required to remain under USAF control until closure date.

The Philippine Air Force/CABCOM (Clark Air Base Command) was responsible for security within their packages after turnover. USAF/DOD security personnel provided point security only on any mission essential facilities remaining within a PAF controlled area.

When the above turnover plan was approved, the 13AF Facilities Drawdown Programming Plan (PPlan) was developed. The PPlan time phased actions were developed to coincide with the turnover of the various packages. Organizations cut back services provided, removed all non-excess equipment items and reduced manning requirements as each package turned over. Also, organizations consolidated remaining services/personnel into the minimum number of facilities possible. The planned result was to have the majority of services terminated and all non-excess equipment removed by mid November.

In determining the status of equipment the following measures of merit were considered. Non-excess equipment had to be critical to DOD needs, accountable, serviceable and had to be economically viable to remove. Computers, weapon system spares, munitions, most communications equipment, special purpose vehicles, high value medical equipment, weapons and mobility/CWDE gear met the non-excess criteria and was shipped from Clark AB. Excess equipment remaining in place for transfer to GOP included most office furniture, some powered/non-powered ground support equipment, general purpose vehicles and general support supplies. All base agencies made inventories of what they determined to be excess equipment. These excess defense articles were consolidated and submitted through HHO for the congressional notification process. The Excess defense article list met the congressional notification period and the USAF was authorized to transfer these articles to the GOP free of charge.

AAFES

AAFES was able to open twelve facilities within three weeks of the eruption with one shoppette opened three days after the eruption. During the final 45 days, MFES operated a small retail facility (50 linear feet) in the corner of the Chambers Hall's Snack Bar. We kept gasoline pumping until 15 Nov, a Barber/Beauty Shop open until 23 Nov and the Video Rental open until 25 Nov 1991.

The average sales on Clark for 1990 was $4.4 million a month. From July through October 1991, sales averaged $400,000 per month.

We used 137 containers to move $18 million in merchandise and equipment to other MFES exchanges in WESTPAC.

A RIF type process lowered the work force from 800 to 200 by August (RIFs were processed in Oct). Remaining personnel were released when no longer required and AAFES terminated operations 25 Nov 91 with 10 personnel on the payroll. The pre-eruption payroll averaged 4 million pesos ($148,000) per month. Severance pay for the 800 personnel was 85 million pesos ($3,148,000).

We feel we've done an outstanding job not only in protecting and recovering our assets, but most importantly providing a much needed service. It goes without saying that we couldn't have done it alone. Three groups deserve a lot of recognition. They are:

1) Our employees. These folks traveled across rain/mud swollen rivers (or walked across) to get to work. They were responsible for getting facilities open, shelves stocked and merchandise shipped.

2) The Navy Exchange at Subic Bay. The exchange at Subic suffered more structural damage than we had at Clark including destruction of all of their warehouses. Regardless, they were most cooperative in shipping merchandise to keep Clark Ash Warriors in stock of common sundries.

3) The Air Force personnel at Clark. Everyone worked together to help us accomplish our mission. From the Communications Group getting phones and all the important computer lines up - to SPs guarding our facilities. Everyone demonstrated the ability to cut through needless red tape. This was a major undertaking and the Air Force did it.

AFOSI

The Operational Track, supervised by Dist 42/DO/IVI but primarily headed by Det 4240/CC, focused on antiterrorism collections, Protective Service Operations and reactive criminal operations.

Det 4240 Operational Team conducted work on over 40 pending cases and opened several cases in addition to conducting numerous Protective Service Operations and continued IIR publication. Operations transferred pending cases and evidence to OSI units, FBI, NIS, USSS, etc., and assisted in preparation of Det 4240 supplies and equipment for shipment. A small task force completed work on 91 42D6-279 and helped SJA prepare for General Court-Martial of subject at Hickam AFB, HI.

The Closure Track, headed by Dist 42/DO, initiated closure actions. Initial actions focused on protection and removal of USAF/AFOSI property and assets left in Bldg 2127, Dist 42 HQ. All assets were removed and shipped to locations identified by HO AFOSI/LI/LI/LG; primarily Dist 46, Yokota AB, Japan by 4 Sep 91. Necessary actions were taken to close Dist 42 HO at CAB and transfer all district responsibilities to Dist 42, US Embassy. Closed case files were transferred to staging area for extensive review and follow-up of command actions. AFOSI also transferred or disposed of over 1000 items of evidence.

CHAPLAIN

Following the decision to close Clark AB, all Catholic and Protestant Chaplain Fund Property was donated to over 20 different Catholic and Protestant Parishes which had been devastated by Mt. Pinatubo. This was accomplished by 30 September and done in accordance with AFR 265-9 and direction from HO PACAF/HC.

The Chaplain Funds were dissolved by 30 Sep 91. Two cashier checks, for the Protestant account #16580998 for $38,065.86 and for the Catholic account #165081000 for $16,431.74 were mailed to HO USAF/HCB per AFR 265-9.

An additional four refund checks were also mailed to USAF/HCB totaling $5,544.00 from United Airlines for tickets not used because of the eruption of Mt. Pinatubo and the cancellation of the Christian Concert. These checks were not deposited into the Protestant account as it was already closed.

The original receipts from the donation of Catholic and Protestant Chaplain Fund Property, along with the Catholic and Protestant Fund records were mailed and are now on file with HO PACAF/HC. Two grand pianos and two Allen organs from Chapel One and Chapel Two were crated and shipped to AFLC.

Chapel One and the Activity Building were turned over to CABCOM on 17 Oct 91. Chapel Two turned over on 15 Nov 91. After this turnover, all remaining Chapel programs/services were conducted at the Airmen's Open Mess or at the Chamber's Hall Chapel Annex through 26 Nov 91.

CIVIC ACTIONS

Following the eruption of Mt. Pinatubo in mid June and the ensuing evacuation from Clark AB, Civic Actions (CA) reconstituted its operations on base on 28 June. The challenges faced were considerable. Prior to the eruption, there were four U.S. personnel assigned; however, two USAF enlisted personnel that supervised the CA warehouse were evacuated and did not return to Clark.

Physical destruction was substantial. The CA warehouse (Bldg 2885) had collapsed and the resulting damage obliterated numerous records and receipts, as well as destroying large quantities of relief supplies and medicines. CA's main office (Bldg 2443) sustained minor damage but remained structurally sound and, once substantial clean up was accomplished, became a combined office and warehouse facility. It took approximately one week to salvage recoverable facilities, clean usable facilities and prepare to undertake substantial relief operations which began on 5 July with provision of tents and sleeping bags to Poroc, Pampanga.

A major challenge for our relief convoys throughout the entire period was bypassing numerous downed bridges and negotiating riverbeds swollen with mud and volcanic debris. Sudden heavy downpours posed the ominous threat of triggering cascading walls of mud that could engulf our operating locations or isolate us from the base. One mobile medical operation to the hard hit municipality of Conception Tarlac had to be cut short and all personnel hastily evacuated due to a fast moving mudslide. Consequently, frequent weather checks and continuous communications with the 3TFW Command Center were necessary to preclude dangerous situations.

Other challenges included a re-orientation of mission focus from generalized nation-building programs throughout five central Luzon provinces to a more specific disaster relief operation in the immediate vicinity of Clark AB. Realizing that the scope of the disaster had overwhelmed central and local government resources, CA began a three pronged program of medical operations and relief goods distribution in dozens of evacuation centers, while providing as much usable excess U.S. Government property as possible local schools, hospitals and municipal offices through the DOD/State Department Humanitarian Assistance Program (HAP).

CA's goal was to create a highly visible USAF profile among the tens of thousands of homeless evacuees huddled around Clark AB by demonstrating the continued concern of the U.S. government for their plight even as a major evacuation of U.S. personnel and valuable equipment from Clark was under way. To the extent that CA was successful in maintaining or enhancing a positive and caring USAF image in the surrounding communities during this period, a direct spin-off would be enhanced security for U.S. personnel and resources, both on and off base. Toward that end, CA's relief distribution continued until just prior to the final Clark AB turn-over ceremony on 26 November.

CA's disaster relief operations around Clark AB included 24 mobile medical operations treating over 14,000 evacuees, and 48 relief goods missions distributing 125,000 individual meals and 3,973 blankets, as well as numerous other excess property contributions.

COMMISSARY

Following the eruption of Mt. Pinatubo, commissary operations virtually came to a halt. The commissary main store, remote warehouse and three of the five buildings located at the cold storage area were left intact. However, the main store was filled with mud and debris. Several thousands of dollars worth of merchandise and equipment was destroyed or damaged beyond repair.

After a clean up of the main facility was completed, a case lot sale was conducted from 22 Aug to 7 Sep 91. The sale accumulated a total of $800,193.51 cash sales to patrons.

To support mission essential personnel (Ash Warriors) manning Clark AB, a commissary Wee-Serve was set up in the store adjacent to the main commissary building. The Wee-Serve began operations 22 Aug 91 and concluded operations 3 Nov 91. Cumulative totals for that period was $263,130.30.

The case lot sale conducted at the main store and Wee-Serve operations was not expected to deplete the remaining merchandise still in the warehouse. Daily transfers were conducted from Clark to Subic Naval Commissary. In excess of 1.1 million dollars of perishable and semi-perishable merchandise was transferred. This amount also includes $14,000 of troop subsistence transferred to the Naval Supply Depot. Items not transferred to the Naval Commissary were used by base Civic Action in helping residents affected by Mt Pinatubo's eruption.

Commissary equipment was affected by the eruption also. Approximately $800,000 of commissary equipment that was salvageable and cost effective to be used at other commissaries or US facilities was transferred to Subic Naval Bay.

The effort put forth by the initial 5 NCOs and 1 Airman, 27 local national employees and continued on by the remaining 5 individuals including 2 TDY personnel kept the base fully supported throughout this difficult period.

COMMUNICATIONS

The 1961st Communications Group was tasked with removing the communications and computer systems supporting Clark Air Base, Wallace Air Station, and Camp O'Donnell. These removals also affected remote microwave radio repeater sites at Tinang, Santo Tomas, Angat, Villamor, and the US Embassy in Manila. The 1961CG created a dedicated staff of four people to complete the planning process and monitor system removals. A 20-person removal team was also organized. To provide removal expertise in the specific areas of heavy radar and antennas, the assistance of a 12-person engineering and installation (El) team was obtained from the 1837E15, Yokota AB, Japan. Removal time lines were established based on mandated drawdown dates, communications requirements supporting base closure efforts, systems threatened with destruction, and systems not required. Most communications-computer systems received only minor damage and were totally recoverable. The only equipment not recovered due to corrosive effects of the ash were the antennas which were all abandoned in placed.

Removal of all Wallace AS communications-computer systems was accomplished by 1961 CG, OL-F personnel with minor support of Clark people. Camp O'Donnell equipment was removed. Both installations were completed by 10 Sep 91, with only minor difficulties.

Two communications areas, the satellite communications facility and the high frequency (HF) radio receiver site, were considered in danger of total destruction by high waters and hot mudflows. The satellite communications area, being closer to the volcano and a river, was the most threatened facility. This facility received major water and mud damage to the solid state uninterruptible power supply area, due to a collapsed roof, but the actual system received only minor water damage and was identified as salvageable. To ensure this $29 million facility was salvaged, the maximum removal time was established as 30 days and a specialized Army team from Ft Huachuca AZ was gained through coordination with Defense Communications Agency (DCA) and PACAF. Eighteen personnel arrived on 15 Aug 91. Local communications personnel coordinated with transportation, civil engineering, and removal team personnel to provide cranes, forklifts and trucks. Complete system removal was finished on 9 Sep 91. Removal of the HF facility was simultaneously accomplished by 1961 CG and EI people without any significant problems.

Minor adjustments to the removal schedule were continuously made throughout the process. All changes resulted in the earlier removal of equipment to reflect loss of missions or to protect equipment from vandalism. Examples were the decision to close the radar approach and control facility (RAPCON) and reduce HF radio and weather operations. Closure of the RAPCON allowed the radar and radio equipment, 10 sea-vans of equipment, to be removed 45 days ahead of the original schedule. This permitted the early release of 15 people. Reducing HF radio and weather operations allowed the shipment of 6 vans of equipment 60 days early, release of 4 people, and saved the equipment from destruction by vandals breaking into our remote areas.

Last day communications needs were identified as worldwide telephone, message, land mobile radio, and limited high frequency radio capabilities. Telephone service was provided by replacing the $7.5 million, DMS 100 switch with a smaller, 600 line, switch previously ordered to support base requirements. Switch cut-over was accomplished without incident on 6 Nov 91. Crates, boxes, and packing materials were pre-positioned for all remaining equipment to facilitate removal on 26 Nov 91. Removal, packing, and disposition of over 31 different communications-computer systems valued in excess of $140 million were accomplished in less than 120 days.

During the Clark Air Base closure process, the 1961 CG provided training to Philippine Air Force personnel. Fifteen PAF personnel received training in the Air Traffic Control tower. This training covered tower procedures and radio operations. Our operations personnel provided training to two people on switchboard operations. Training on switch maintenance, cable repair, land mobile radio repair, and technical control repair was provided to 30 people by our maintenance branch.

6922 ELECTRONIC SECURITY SQUADRON

All major actions were completed prior to the implementation of the PPlan. Squadron equipment was removed by an ESC team of technicians and shipped to other command units or to depot at Kelly AFB. All classified material was destroyed or shipped to a sister squadron at Kadena AB, Japan. A small mission operation remained at Clark AB until after the closure ceremony. The commander, two operators and a maintenance technician remained until the final day with the remaining equipment driven to Cub Pt NAS and hand receipted to an USMC unit for long term loan.

COMPTROLLER

Upon notification of the closure of Clark, we immediately contacted all required agencies --HO PACAF/AC, Air Force Audit Agency, Defense Finance and Accounting Center-Denver and PACOM central funding. We developed these objectives:

Full pay/travel service to Clark team and evacuees;
Funding for services/supplies for our withdrawal;
Closure and transfer of appropriated accounts.

The next task was to determine which personnel would remain for the closure. We were able to keep a good mix of comptroller personnel which allowed for an orderly closure without interruption of normal day-to-day financial services.

Although accounting became a nightmare because all records were lost during the eruption, a comprehensive, yet simple, manual accounting system was established until computer connectivity was established with Kadena AB. We then coordinated with HO PACAF/AC to determine where our operation would be relocated until the deactivation of the AFO.

A major task to be accomplished was payment of final, severance and leave pay to all the local national and U.S. civilians that were being RIF'd. This required close coordination with MSSQ, CCPO, the local union and our Civilian Pay section. Although these payments had to be computed manually and became a very tedious and time consuming task, we were able to pay all employees prior to 22 Nov 91.

Evacuee Safe Haven benefits were changed four times, each change being at variance with the Joint Federal Travel Regulations. This caused a lot of confusion worldwide and a tremendous workload on our specialists in the form of calls to hundreds of disbursing offices to make sure they had the "straight" word. In the end, AFMPC/DPMA msg 202200Z Sep 91 laid out clear and final guidance on evacuee benefits.

There was significant posturing from HHQ on how to report costs associated with Pinatubo and subsequent withdrawal. We tracked the costs carefully and wound up with the following:

COMMODITY	*FY 91 ($000)*	*FY 92 ($000)*
Civilian Pay	$ 5,719	$ 4,661
Civilian PCS	295	350
TDY Travel	1,475	250
Transportation	1,892	1,407
CE Services	1,281	473
Supplies	1,665	360
Equipment	-0-	3
Miscellaneous	3,326	1,267
TOTAL	*$ 15,653*	*$ 8,771*

On 1 Nov all accounting records were packed up and shipped to Kadena AB where the AFO and four enlisted people set up a small finance operation for the purpose of final balancing and closure of the AFO account. Estimated completion date is May 92.

CONTRACTING

The closure process for the Contracting Center presented great challenges to even the most experienced contracting personnel. Numerous milestones and checklists had to be developed to ensure an orderly and dignified shutdown of the center. Plagued with poor communication lines, language barriers, time zone differences and a geographic location where the supply lines are very long, the center initiated the closure process with local and distant contractors.

The initial task for the center was to assess the actual workload and determine the proper course of action to cancel and/or terminate all existing active contracts. It was discovered that over 6,600 small supply and service purchase orders valued at $12 million and 84 complex construction and service contracts valued at over $64 million had to be canceled. In an attempt to expedite this compressed process, cancellation letters were sent to all vendors, both local and those in the United States. The letter informed vendors that all orders that had not been shipped were being canceled, and if the vendor concurred, his signature was required and requested to return the letter. Vendors who had questions or did not agree with this cancellation were given 20 days to contact this office for resolution. This process proved to be extremely effective as over 95% of these purchase orders were canceled using bilateral agreement, saving the US government thousands of dollars in possible claims.

The termination process for the 84 construction and service contracts, each over $25,000, was more difficult and time consuming based on the complexity and the regulations governing these contracts. The initial attempt was to "Terminate for Convenience of the Government" the majority of these contracts. However, after proper study, it was discovered that this process would be extremely lengthy and prolong the termination process by an additional year. It was decided to use deductive modifications, whereby remaining work on contracts would be deductive and with the contractor paid for work completed and cost incurred. This process alleviated a possible one year settlement and allowed most contracts to close out in less than 30 days.

Remaining contracts that require Termination for Convenience or other contractual actions such as claims and Armed Service Board of Contract Appeal cases that require lengthy settlement periods was transferred to the two man contracting office in Manila. This office will remain open until June 1992 with Andersen AFB, Guam, assuming administrative duties.

DRMO

Clark AB DRMO facilities experienced heavy damage from the eruption of Mt. Pinatubo. All scrap and other items stored outside were covered with ash and mud. Additionally, warehouses suffered interior damage due to flooding and mudflows. After cleanup operations, DRMO prepared for a large influx of excess property. Also, in excess of 800 POV's were turned into DRMO as Government Claims Property. To meet the closure date an "expedited method of negotiated sales" was used. This enabled DRMO to successfully sell all excess property (including POV's) and obtain sales in excess of $450,000.

ENGINEERING

From an engineering prospective, Clark phase-out operations have challenged us in six major areas:

1. Packout of equipment
2. Continued operation of utility systems
3. Resolution of environmental problems
4. Inspection and turnover of facilities
5. Preparation for unexpected natural disasters
6. Finalization of evacuation-related housing issues

Equipment packout went very smoothly with our engineers continually being among the leaders in beating wing packout schedule milestones. We initially made comprehensive surveys with a team from 3d Supply Squadron to determine material and equipment for packout, identify property as excess, and determine what material was immediately available for our counterparts at Subic, Cub Point, the U.S. Embassy, and Diego Garcia, among others. Once we determined what these agencies could retrieve, they very efficiently loaded and hauled the material, under our supervision, with very little impact on our resources. These appliances varied from new appliances (dishwashers, ranges, refrigerators, washers, and dryers) to refrigerators in abandoned quarters full of rancid food. We were able to ship 3000 new appliances (nearly 70 SeaLand vans) to Andersen AB.

A contractor was hired to clean refrigerators in quarters at a small fee to abate environmental problems, and the remainder of appliances, nearly all used or damaged, were either centrally stored or left in place.

Our people--structural, heavy equipment--were instrumental in supporting extensive material extrication operations from downed facilities base wide. Over $750 million in equipment and supplies were successfully retrieved.

Operating our utility systems was a real challenge that our technical experts handled brilliantly. With the continual presence of the abrasive, corrosive ash, mechanical systems failed prematurely. Backup parts were not available. Operators were forced to remove parts from inoperative systems and secure emergency repairs from local vendors.

One problem that complicated our efforts was the deterioration of facilities after the turnover to the PAF. At times, plumbing fixtures were removed from their mounts leaving water to be lost. As a result, we secured entire areas of electricity and water service to continue support to remaining U.S. personnel. The systems (water and electrical) were extremely fragile and conservation was a top priority. The likelihood of bringing these systems back to life is small--repairs would be extensive and costly, breaks would be hard to find as several inches of ash cover most valves, and exposed components deteriorated rapidly from corrosion.

Throughout, we encouraged the PAF to work along with our experts both during formal seminars and training sessions as well as OJT. Response was sparse. Most of the PAF individuals who were assigned to us had little or no prior experience or at best were generalists rather than specialists. Our systems are extremely complex. Continued operation after the U.S. departure will be difficult.

Our Liquid Fuels Maintenance people supported complicated fuel transfer operations to facilitate removal of over 2.2 million gallons of POL products from the tank farm prior to turnover.

Environmental cleanup operations were directed at the removal of stored hazardous waste and PCB from stored transformers. Since we lacked the manpower and contracting capability to remove this material before the departure of U.S. forces, we contracted this operation through the Public Works Center at Subic. Operations were completed by mid November.

Facility turnover, despite the scope of the task--nearly 3800 facilities--went very smoothly. Simplistically, organizations cleaned and turned facilities over to CE, CE and the PAF performed a joint inspection of the facilities, and then facilities, in phases, were signed over to the PAF with a simplified administrative procedure. Two major problems were encountered especially in the housing areas, both related to access. First, hundreds of houses were relatively inaccessible--debris, trees, mudflows and ash made vehicle movement nearly impossible. Employing in-house and contract equipment and labor, we ensured that access was gained. Secondly, hundreds of evacuees left with their keys. Our locksmiths had to forcefully enter over 3000 facilities and then replace the locksets in most cases. This forced entry procedure was necessary due to the extremely ambitious packout schedules and the need to gain access to hundreds of facilities each day.

One of our major challenges, especially during the rainy season, was to prepare for flooding, ash falls, mud flows, or wind damage from either torrential rains or typhoons. We had to hire a large labor force to hand-dig drainage systems, employ contracted heavy equipment to dredge waterways, and carry off huge quantities of ash to preclude it from ending up in drainage systems. As rains came and went, we needed to repeat this procedure at least four times, as systems would fill time and time again. Requirements contracts for both labor and equipment saved the day. Nearly $800K was spent to ensure that we were properly protected. Throughout, recovery and prevention were our aims rather than reconstitution or reconstruction.

Our housing office was initially involved heavily in ascertaining the status of evacuees and identifying quarters off-base that needed to be packed out. They were also involved as key members of assessment teams, inspecting quarters and recording damage to evacuees'belongings. Finally, they geared up to support claims from Air Force homeowners downtown who either lost their homes or who are unable to sell their homes. We needed DOD-level officials to visit Clark to administer the Homeowner Assistance Program (Public Law 89-754, Section 1013). Despite our requests and those of PACAF/DE, we were unable to receive the on-site assistance.

Despite our dwindling resources--equipment, supplies, and people--the final analysis will show that our engineers and firefighters did a remarkable job in protecting assets, keeping systems on line, and getting high-value materials and equipment on the way to other users. Our people were extremely professional enthusiastic, and resourceful in not only handling the ordinary but also in attacking and resolving the extraordinary.

FAMILY SUPPORT CENTER

The personal effects of personnel evacuated to CONUS were mailed to appropriate forwarding addresses. Remaining FSC equipment/property was transferred to the FSC at Naval Station Subic Bay.

Air Force Aid was provided to remaining personnel as needed. All Air Force Aid Society files were either forwarded to the gaining station of the member, or to HO AFAS. All AFAS records were audited prior to shipment to AFAS. All other AFAS property to include files, canceled checks, check register was mailed to HO AFAS.

Remaining unused AFAS checks were, at the direction of 3CSG/CC, shredded with two witnesses to the action.

Limited counseling was provided to individuals in need of short term counseling through 22 Nov 91.

HOSPITAL

Upon returning from Subic, HO PACAF/SG was contacted and concurred to zero all Medical Center supply and equipment accounts. This made it easier to add items back to our accounts rather than trying to find the items and adjust the inventory during salvage operations.

Further coordination with PACAF/SG on a single destination for all 23 million dollars worth of salvaged medical and dental equipment simplified the transportation requirements. The timely arrival of requested medical maintenance and medical logistics TDY personnel provided us the needed manpower to complete the salvage and excess inventory operations.

Due to the lack of power to operate the hospital's elevators, many bulky and expensive medical equipment items were initially stranded on the 2nd floor (Surgery and OB). By using two 4k forklifts and a borrowed crane we were able to salvage these items.

Because of the size of the Medical Center, surrounding buildings and having ongoing medical and dental operations at an entirely different part of the base, good communications was vital. The Motorola STx-800 handheld radios with a separate medical radio net worked extremely well and enhanced our overall operation.

Finally, remaining medical clinic personnel were able to provide essential services throughout the drawdown. Basic medical care was provided to all qualified personnel during the majority of the drawdown, with some specialties available until late October.

JUDGE ADVOCATE

International Law: We began the closeout period with seventeen "international hold" cases. All cases were closed out.

Military Justice: We began the closeout period with three general court martials. All cases were successfully tried and concluded. In addition, over 20 Article 15 actions were completed.

Civil Law: Legal assistance was provided to the base population until 15 October 1991. Powers of Attorney were available until 25 November 1991. A Joint U.S./Government of the Philippines program to sell tax exempt privately owned vehicles (POVs) was implemented. A 50% ash-damage reduction on the taxes assessed on the vehicles was successfully negotiated. A variety of quality force issues was resolved including two administrative discharge boards under AFR 39-10. One officer discharge case under AFR 36-2 was transferred to the 633ABW/JA at Andersen AFB, Guam. In addition, a variety of other issues associated with the withdrawal of American forces were resolved to include: closing down 38 private organizations, disposal of MWR property, reviewing all contracts for termination and disposal of two Merit System Protection Board cases.

Claims: Over 840 claims were processed with over $650,000 paid for damages resulting from the eruption of Mt. Pinatubo. Over 1,600 POVs and over 5,000 residences on and off base were inspected for damaged property.

Administration: Office files and equipment were transferred to Andersen AFB, Guam as well as all remaining unresolved issues.

MAINTENANCE

The workload confronting maintenance upon return from evacuation was mammoth. Only 28 permanent party 3TFW maintainers were left at Clark to accomplish the out-load of munitions storage and aircraft maintenance while simultaneously supporting the pack-ups/shipments of household goods and automobiles for over 1,000 maintainers and their families. We immediately prepared a detailed milestone chart and assessed TDY manpower requirements --45 TDY personnel were requested and received in July and August 1991 to work the following:

1. Pack-up and shipment of 298 truckloads of munitions and related equipment valued at over $100 million.
2. Closure of 73 munitions custody accounts from Wallace Air Station, Camp O'Donnell, Crow Valley Range, and Clark Air Base.
3. Salvage and shipment of $40 million of jet engines and related support equipment from collapsed buildings.
4. Salvage and shipment of $11 million of WRM armament pylons and equipment from collapsed buildings.
5. Salvage and shipment of $4.5 million of precision measurement laboratory equipment.
6. Salvage and shipment of over $55 million of ground support equipment from AGS, CR5, and EMS shops.

All of the above actions were completed by 20 Oct 91 - weeks ahead of schedule. In November, only 24 maintenance personnel (including contractors) remained to handle transient alert, closure of helicopter operations (last flights on 26 Nov) and final munitions accountability tabulations at Subic Naval Magazine. The three UH-1 Ns will be prepared for shipment and transported from NAS Cubi Pt to Alaska in early December.

MANPOWER AND ORGANIZATION

HQ PACAF/XPM at Hickam AFB, Hawaii was responsible for the manpower actions associated with the closure of Clark Air Base. Returning from Subic the Management Engineering Team (M~ co-located with 13AF/XP to work the manpower issues. The most important issue was the development of the new 13AF manpower requirements, which were transferred to Andersen AFB, Guam. During this period, the 13AF/DP created the Manpower and Personnel Readiness Center (MPRC). This was composed of two manpower and two personnel folks. It's mission was the accountability of ah ash warriors (U.S. Military, DOD Civilians, Contractors, etc.) and TDY manning assistance requests.

MISSION SUPPORT SQUADRON

1. Information Management:

A. Reprographics: A total of $171,296 worth of reprographic/duplicating equipment was disposed of lAW AF regulations and coordination with HO PACAF/IM. TDY assistance was brought in to break down equipment and to send it to various units within PACAF. Each gaining organization paid the shipping costs. Each piece of equipment was assessed for depreciation value, wear and tear, and damage due to water and the corrosive ash. A total of $128,690 worth of equipment was shipped. The remaining items valued at $42,606 were turned over to DRMO for bidding. This included two offset presses and a plate maker.

B. Records: Due to the severe environmental factors existing immediately following the eruption and the potential for further damage due to rain, ash and mudflows, a plan was developed to evacuate critical records. Taking into consideration the airlift available and the traveling distance, Kadena AB was chosen to stage and/or make final disposition of records. The priority one records were medical, dental, personnel (military), education records (DODDS) and civilian personnel records. A total of 24 pallets were shipped in an eight-day time frame. All other base records were shipped lAW AFR 12-50. Items not determined cost effective to ship were shredded through the MWR Recycling Center

C. Publication Distribution Office: The building which housed the PDO was condemned by Civil Engineering; power and water were never restored. As a result, very few forms were able to be salvaged. Any excess forms on base were shredded through MWR recycling lAW AFR 4-71, para 7-1 as it was not feasible to transfer unserviceable forms and publications. All form requests were made through PACAF/IMP. Lateral support was handled through the PDO at Kadena AB. All accountable forms were disposed of lAW AFR 4-71, para 5-7b. All associated documentation (AF Form 145, Certificate of Destruction, AF Form 213, Receipt for Accountable Form, and AF Form 505, Accountable Forms Stock Control Record) was forwarded to the designated staging area.

D. Copier Program: Aggressive actions were taken to retrieve all copiers for turn in to the contractor. A few copiers were unable to be retrieved due to the unsafe condition or excessive structural damage to its facility. Any copier unable to be retrieved or lost, was documented and coordinated through contracting and the Xerox contractors. Out of 202 copiers, seven were lost or unable to be retrieved, and 195 were recovered.

2. Civilian Personnel: Separations/Reductions-in-force actions were conducted as they were and when they were to confirm that the United States Government and USAF were compassionate employers. Effective dates were chosen to give the employees the greatest benefits or, conversely, not to separate them a few days before a Collective Bargaining Agreement benefit was due (effective 1 Oct91).

3. Social Actions: All documentation was disposed of lAW AFR 12-50 and AFR

30-2. All active files, historical records, films, videos and audio cassettes were sent to HO PACAF/DPAH. All actions were completed by 11 Sep 91

4. History: TDY support from HO PACAF/HO and 313 AD/HO completed all actions necessary to properly dispose of files, classified and memorabilia. All 3TFW memorabilia was sent to 11 AF/XP. The remaining items from the Fort Stotsenburg museum and the history offices were sent to the main office at Maxwell A~B. The final historical report will be handled by HO PACAF/HO.

5. Military Personnel:

A. Mission Essential List: A personnel roster was designed to ensure accountability of the Clark population after the evacuation. It was used to track arrivals and departures. This list was utilized in the absence of the PDS.

B. Assignment Action Team: This team comprised members from HO AFM PC, HO PACAF and HO MAC. They were requested to ensure all mission essential personnel received assignments and orders in hand in the event another evacuation was necessary. The team provided on-the-spot assignments with a high level of satisfaction. By having all personnel on assignment, it facilitated the shipment of automobiles and household goods and allowed the movement of dependents evacuated to move to their PCS location.

C. Awards and Decorations: End-of-tour (E~ and Fiery Vigil decorations were mass processed. Commanders made recommendations for the level of award which were approved by the decoration authority (13AF/CC or 3TFW/CC). Certificates and citations were typed by the units and then processed. In excess of 4000 decorations were processed.

D. Records: A five person team was sent to Kadena AB, Japan with the Unit Personnel Record Group (UPRG), medical and dental records to ensure the records'survival and to forward records to gaining bases. If the end assignment/location was unknown, the records were sent to HQ PACAF/DPA. All 353SOW and 624MA5G records were left at Kadena because these units'servicing CBPO was redesignated to that location. Remaining medical and dental records were transferred to the Kadena Clinic for final disposition.

The 3d Mission Support Squadron (3d MSSQ), had an enormous task-- one unlike any other mission support squadron has ever encountered. Up front, the requirements were uncertain, the hours were long, there were many errors; but, fortunately, the men and women of the 3d MSSO performed exceedingly well, which allowed this base to close ahead of schedule. The primary function of the 3d MSSQ was to run all of the military and civilian personnel programs, and manage the Information Management process for the base. Our biggest challenge was to account for people--knowing where our people were, when they had to leave, where they were going, getting them moved to the right places (and often that translated to a new building or shelter), and reporting this information--accurately--to our leadership.

We did not have a computer system. The Advanced Personnel Data System (PDS) couldn't be used and this system was our mainstay in tracking and moving people. Unfortunately, by evacuating thousands of people our PDS became unreliable and we did not have the time nor the talent to fix it. As an alternative, we built a database using a 10 Jun 91 PDS data file, and then kept using this database on a PC. We would channel rosters by unit to the unit commanders and then the commanders reported to us changes in strength. The ultimate "headcount" became the only way to track people. All other personnel management actions were dictated by this database which became known as the Mission Essential List. It became widely known as our "Bible" and without it we would have surely failed.

If there was any lesson learned about surviving--in a personnel sense--during this natural disaster and subsequent rapid closure, it was accountability. From dropping assignments and port calls on each Ash Warrior, to notifying them of promotions, to tracking them down on the weekend to advise them of a death in the family, a personnel agency had to have a handle on where the people were.

MORALE, WELFARE and RECREATION

In accordance with PACAF directives, MWR equipment that was considered to be in good condition and was requested by other PACAF bases was packaged and shipped. Guam was the primary port with some items forwarded directly to other PACAF bases. Items shipped included such things as athletic equipment, childcare Center equipment, Golf equipment, Open Mess equipment, paper products and other recreational items. Other items not suitable for shipment were sold or transferred to Subic Bay Naval Station.

Selected MWR assets were put up for sealed bids: NAF horses, kitchen equipment, bowling equipment and supplies, Arts and Crafts equipment and supplies, airplanes, MWR vehicles and other related NAF equipment. The bid openings were observed by the legal office and NAFFMB personnel.

MWR was tasked to contact private organization officers to dissolve all private organizations at Clark. This required considerable time and effort to locate the officers, get dissolution letters submitted and close out each account legally.

A consolidated sales store was created to sell excess resale inventories and NAF property. It was determined that it would be more cost effective to sell the goods, even at reduced prices, than to attempt to redistribute the items to other Air Force bases. To redistribute these goods, we would have had to determine where the items were needed, possibly split the inventories to fill the requirements, pay to have the items packaged and shipped and chance spoilage, loss or damage during shipment and handling. Items were consolidated into one account and sold through the consolidated sales store beginning on 16 Jul 91. Prices on resale items ranged from "cost" to "50 percent of cost," depending on the type of item, age, condition, etc. Limited quantities of NAF property from billeting activities were also included in the sales. The sales store remained open until 15 Nov 91, with $381,000 collected.

Following "Fiery Vigil," many MWR NAF employees were brought back for duty. Some were used to clean up and reopen facilities (i.e., The Airmen's Club, Golf Course and Jose's, tennis courts, Liberty Gym, etc.), and others were used to clear debris, stage equipment to be shipped and protect facilities. In all about 600 employees worked between July and October. Effective 1 Nov 91, all employees were terminated and on 2 Nov 91 approximately 450 MWR NAF employees were rehired on a temporary intermittent status. These employees were used to operate club, golf, gymnasium, etc., facilities, accomplish the final payroll and financial statement closeout and complete final pack-up and shipping of equipment. All employees received their final pay, separation pay and bonuses by 22 Nov 91.

OPERATIONS

Upon completion of Operation Fiery Vigil, the mission of the 3d TFW Operations Division was to support the 3d TFW Mission Objectives by accomplishing the following actions:

1. Conduct safe helicopter operations with three assigned UH-1 Ns
2. Continue base operations and command post functions
3. Assess damages to 3 TFW/DO and 6200 TFTG facilities and equipment
4. Protect remaining assets
5. Pack and ship household goods for assigned personnel
6. Pack and ship/turn in mission essential equipment
7. Train PAF operations personnel in base operations functions
8. Turn over assigned facilities to PAF

All actions were successfully completed. The following organizations terminated operations at Clark AB on or before 26 Nov 91:

3d TFW/DO: 3d TFW/OTF/DOC/OTM, 3d TFS, 90th TFS.

6200 TFTG: 6200 TFTS, 1st TESTS, 3d TEWTS.

Memorabilia for the 3TFW and the 9OTFS were sent to Elmendorf AFB, and 62OOTFTG's were sent to Eielson AFB.

Helicopter Ops: Three UH-1 Ns flew over 350 successful sorties in support of the US Geological Survey (volcanologists), VIPs, 13 AF/CC, mission support (medevac), and aircrew training. Superb helicopter maintenance by Kay Associates resulted in a 99% in-commission rate. Helicopter Operations terminated on 26 Nov 91 with a final sortie flying 13 AF/CC and 3 TFW/CC from Clark AB to NAS Cubi Pt.

Command Post: During the close down phase at Clark AB, the Command Post mission was to coordinate/compile data (SITREP, POV issues, household goods), to provide liaison with staff agencies, and to host distinguished visitors to the briefing area. Operations terminated on 26 Nov 91.

Base Operations: Base operations trained PAF personnel and provided augmentation personnel to successfully complete the shipment of HHGs at Clark AB and POVs at Naval Station Subic Bay. USAF base operations terminated 22 Nov 91.

3 TFS/9OTFS: All HHGs were shipped for evacuated personnel and recoverable mission essential equipment was turned in to supply.

6200 TFTG: The LORAL (Ford) Aerospace Crow Valley Range contract was terminated on 30 Aug 91 after the successful recovery, pack-out, and shipment of over 200 million dollars worth of high value equipment (threat emitters, radars, vehicles, etc.). Contracts for drone and tracking operations at Wallace AS were terminated effective 30 Sep91.

6200 TFTS: Conducted a unit move to Alaska and transferred all mission essential equipment to Eielson AFB and Elmendorf AFB for use in conducting Cope Thunder-type large force training exercises.

1 TESTS: Terminated Combat Sage drone operations at Wallace AS and shipped assets lAW HQ PACAF/DOO direction. Inactivated 1 TESTS 30 Sep 91.

3d TEWTS: Successfully closed Camp O'Donnell AS and transferred the facility to the P~F on 16 Sep 91. Squadron inactivated 1 Oct 91.

PLANS

Agreements: Once notified of Clark's withdrawal and closure, requested waiver to the 180 day termination notification required by AFR 11-4 and DOD 4000.1 9R. Approval received from HO PACAF/LGX 142050Z Aug 91. Cancellation notification and required documents (DD Form 1144/AF Form 149) were distributed Sep91 to the suppliers/receivers of all current Interservice Support Agreements, Host-Tenant Support Agreements and Memorandum of Understandings for action. Updated the Support Agreement Analysis Program Report upon receipt of signed DD Form 11 44/AF Form 149 and forwarded the final report and wing agreements files to HO PACAF/LGX, 15 Nov 91. Forwarded a processing instruction letter to all units and any final closure correspondence will be forwarded to HO PACAF/LGX after closure of Clark AB.

War Reserve Material Program: Directed the inventory, recovery and shipment of over 7500 War Reserve Material assets. Despite the evacuation of nearly all of the account custodians, the inability to produce automated account documentation and the collapse of the warehouses and storage areas, 97 percent of this critical combat capability value at over $265 million was recovered. Final disposition of these assets was provided in the 29 Aug 91 Clark AB War Reserve Material (WRM) Program Closure Report. Due to the need to quickly recover and ship these items, most assets were shipped to Andersen AFB, Guam, for subsequent redistribution. The WRM Closure Report serves as the official written document for the closure of Clark AB's WRM program, and was forwarded to HO PACAF/LGX for their information/action.

Operations Plans: All 3TFW classified plans were rescinded 27 Sep 91. Remaining 3TFW plans were rescinded effective 26 Nov 91. All Top Secret documents accounted to 3TFW/DOX were destroyed 6 Jul 91.

POSTAL

The Clark AB postal service centers performed their duties in an excellent manner during the base closure period. With the initial "Fiery Vigil" evacuation to Subic, the staff immediately set up postal facilities at Subic so that the evacuated families would continue to receive full mail service.

Upon the unit's return to Clark AB, 70 local national employees were hired to assist the remaining postal workers in all aspects of postal services. Initial backlog of mail took over a month to process with extensive community volunteer support.

Each member of the "mission essential" team was given a new PSC box number at PSC #2 to speed delivery of their mail. Although there were some initial problems with the change of PSC's and box numbers (this coincided with a worldwide APO number change) all problems were worked out individually.

During the three months following the eruption the main APO accepted 470,390 pounds of mail with meter sales of $508,083.00. This was due to the TMO office authorizing the "mission essential" staff priority mailing of household goods. This action increased meter sales by a factor of five. Full postal operations remained in effect until 25 Nov 1991. The Manila APO has agreed to forward all incoming Clark AB mail after 26 Nov 91 lAW addressee's change of address instructions.

PUBLIC AFFAIRS

In the five months following the eruption of Mount Pinatubo and the decision to withdraw from Clark, Public Affairs escorted hundreds of national and international media. They also gave hundreds of interviews, briefed community leaders on the withdrawal, developed and completed a plan to turn in equipment/drawdown the office. Additional taskings included printing more than 100 daily *Philippine Flyers*, and completing end of tour reports and decorations for current and departed office members. Public Affairs did the narration and escorted media for the turnover ceremony on 26 Nov 91.

SECURITY

The 3d Security Police Group (SPG) faced the challenge of reducing its personnel and equipment while maintaining the security integrity of the base. This unique protection/attrition dilemma required a non-standard solution. Before releasing any resources, the Group identified comprehensive base protection requirements through 26 Nov 1991. This was necessary because of Clark Air Base's history of having the highest criminal statistics of any base in the U.S. Air Force.

When magnified by the Mt Pinatubo disaster and severe economic devastation, local national intruders became desperate and considerably more aggressive. The solution chosen, to drawdown security police personnel and equipment while continuing a high degree of security, required two plans.

The first plan enabled the 3d SPG to fully implement its Air Base Ground Defense (ABGD) plan for the installation. The base was sectorized, troops were assigned to permanent areas of operation within the sectors, and an overriding emphasis was placed on detection through the use of tactical sensors, night vision equipment, trip flares, military working dogs, military horse patrol teams, and the continued use of "CYCLOPS"--a PAVE TACK infrared pod mounted on top of the control tower. Philippine Air Force Security Police, Philippine National Police and Philippine Army units were closely integrated into the defense system providing better security for Clark AB and its personnel. The ABGD structure permitted a flexible protection concept that could adapt quickly as intruders changed their tactics. The results were unprecedented with successful thefts cut to one third of the pre-eruption rates. This was accomplished while the number of attempted intrusions soared to between 5 to 10 times previous levels. The 3 SPG detected over 2,000 intruders, apprehended over 350 offenders and recovered property in over 140 instances worth in excess of 25 million dollars.

The second plan determined the reduction rate of SP personnel by tying port call dates to the turn-over of facilities to the Philippine Air Force. Critical security police equipment was also maintained with specific security requirements in mind. Again, the results were positive. Over 980 SP personnel and 10 million dollars worth of equipment were shipped out within established critical time criteria.

SERVICES

The 3d Services Squadron had numerous dissolution and structuring decisions to make during the closure of US operations at Clark AB. We had to determine the disposition of various Non-Appropriated Fund (NAF) properties. NAF large appliances and furniture were turned over to DRMO or transferred to US Government agencies remaining in the Philippines. Small NAF appliances, supplies and furniture were sold to the remaining mission essential force. Property in Chambers Hall (televisions, video players, etc.) was left in place to provide comfortable living quarters for personnel remaining until 26 Nov 91.

Approximately 1 million dollars of Cebu furniture not yet shipped to Clark was shipped directly to Guam. Also, seventeen Sea-Land containers of this furniture were shipped from Clark housing supply to Guam. The Cebu furniture in the new dormitories was left in place, although it was offered to the US Navy.

Computer components (SIMS) for automated billeting, food service and housing supply, valued at $800,000 were shipped to Guam for distribution within PACAF.

Dining Facility (appropriated funds) equipment was left in place or transferred to US units remaining in the Philippines, with the exception of relatively new, brand new or high value items such as reach-in refrigerators and deep fat fryers. These exceptions were shipped from Clark for further disposition within PACAF.

All mortuary equipment, transfer cases and assorted chemicals were shipped to Kadena AB.

One Sea-Land container with four DV (Distinguished Visitor) suites of furniture and appliances was shipped to Guam. Another container of Thomasville furniture, hide-away beds and linens was also shipped to Guam.

Seven Sea-Land containers of laundry equipment and supplies were shipped to Kadena AB. Remaining equipment was turned over to DRMO.

Squadron personnel performed their jobs in an exemplary manner. They went the extra mile to ensure remaining forces at Clark had suitable quarters, nutritious meals and furniture for their quarters. The military and civilian employees within the 3d Services Squadron can be proud of their accomplishments.

NOTE: "Cebu" furniture refers to local purchase furniture procured by contract from Cebu City, Philippines. The contract existed prior to the eruption of Mt. Pinatubo, but had only been partially fulfilled.

SUPPLY

The 3d Supply squadron overcame severe limitations in the Clark AB drawdown. Challenges included the loss of computer operations, collapse of five major warehouses, loss of the main administrative building, mudflows which threatened to isolate or partially destroy the Hill Storage Area and its bulk fuel tanks and a sharp reduction in military manning.

Planning was the first step towards meeting the 120-day drawdown deadline. To ensure the expeditious shipment of high value equipment and recovered warehouse inventory, a decision was made to use a manual system for supply transactions and not to try to resume normal computer processing. This manual system would maintain an auditable transaction trail with documentation carefully controlled and filed. The normal supply squadron organization was abandoned and reorganized by closure function to emphasize that closure was now the primary mission.

The squadron relocated its administrative functions and converted remaining facilities for new uses. To offset the shortage of military personnel, qualified local national employees were rehired and a 26-person Rapid Area Distribution Support (RADS) team arrived to aid both supply and transportation. Recovery tapes processed just prior to the volcano evacuation were uploaded at Kadena AB to reestablish the database. This database was initially used to generate reports and later used by squadron personnel sent to Kadena to adjust financial and supply records.

With the squadron reorganized and accountability procedures were set, initial focus was on three primary areas: first, continued mission support, from MICAP parts to sandbags; second, recovery of supply warehouse stocks; and third, the identification, turn-in and shipment of high value organizational equipment. Recovery of fuel from the hill storage area was not initially planned due to concerns for personnel safety. The fuels drawdown was also a success. Cryogenics plants were dismantled and shipped for use within PACAF. In October, when the PACAF/LG staff recommended the recovery of JP-8 and MOGAS from the Hill Storage Area, over 2.2 million gallons of fuel worth over $2 million were moved by contractor operated trucks to Naval Station Subic Bay.

The warehouse recovery effort was an unqualified success. The most optimistic early estimates were that 50 to 70 percent of the pre-eruption inventory might be recovered. The squadrons warehouse team, aided greatly by the TDY RADS team, recovered an estimated 90 percent ($191 million).

The equipment turn-in program was also successful. Units were given copies of their accountable equipment listings (CA/CRLs) and instructed to identify items as either high value/critical and worthy of shipment or as low value/bulky to be left in place. By establishing schedules for initially turning in non-mission essential equipment, the turn-in team was able to meet the 120-day deadline while recovering approximately $145 million worth of high value equipment. Items left in place were expendable items and low value/bulky accountable equipment that were included as part of the excess defense articles package given to the Philippine government.

TRANSPORTATION

Upon return from the evacuation site, we found severe destruction. The new TMO building had collapsed, files and equipment were destroyed, the runway was closed and would remain closed, the HHG agents off-base facilities and equipment were destroyed, bridges connecting Clark and Subic were collapsed, the roads on-base in the housing areas were frequently impassable and homes were inaccessible due to mud flow around the doors, etc.

In order to expedite the packout of all personal property, we made available facilities and equipment to the HHG agents to include line haul of seavans. At first, we had them sharing warehouse space. Eventually we were able to have each agent in his or her own area. We did this by assigning each contractor a USAF aircraft hangar. As we found the extra space for them, their capacity to pick-up increased. We also coordinated closely with them to ensure they would not run out of packing materials. We opened a dorm to the agents to house their people in so they could get more working hours into the day. All personal property left the base in Type II containers inside of sealed seavans. We sent survey teams consisting of lawyers and TMO personnel to each house to determine who lived there, how much was damaged, how much could be shipped, etc.

We were concerned that the Navy would not be able to support both theirs and our movement requirements. In order to provide expeditious movement of personal property, we decided to ship it all code 4, because the port of Manila had a higher capacity. We knew we would have tremendous numbers of containers of government property that would move out of Subic and we didn't want to bottle-neck their system. Also, the road to Subic was not guaranteed to stay open; Subic had suffered volcano damage and their single crane to load ships was questionable.

In order to allow HHG agents to focus on their effort and gain efficiency, we as-signed specific areas to each agent. Because 2/3 of the base had evacuated, we had "house sitters" who stayed with the movers so there was a blue suit presence in the home, representing the member. We built a plan centered on security and flood/lahar damage. We moved the higher risk areas first and moved strictly IAW the area concept, rank was irrelevant.

When it was decided to evacuate all the personal property from homes we decided to go commercial vice bring in TDY military or pack the homes out with local GI labor. When Panama was evacuated they used GI's and sustained a very high breakage rate. Because personal property is so important to the member, we decided to go commercial and help the commercial industry in every way we could to get their capacity up and provide a quality move for our members. Also, the base infrastructure was inadequate for an influx of additional people.

In order to benefit from concentration of effort and maintain physical security for our customs personnel and house sitters we decided to ship out all on-base HHG before working off-base. The roads to the subdivisions were not clear, bridges were out, the rainy season was upon us and it made sense to finish those areas we had absolute control over while the details of gaining access to off-base homes was worked out.

Abandoned POV's were treated as a claim issue vice forced shipment. The vehicles had sustained major damage and the vehicles which were at Subic were being force evacuated because they were interfering with the Subic clean up. We would not ship unless the member asked. Transportation did not want to spend over $2000 to ship "junk"-- cars valued much less than transportation cost.

Passenger movement was done Cat Y and Cat Z. Cat 'Y'and Cat 'Z'seats were much cheaper than using the contract CAT 'B'flight. The Navy used our normal allocations on the Cat 'B'mission. This way we were able to provide a higher level of customer service at a lower cost to the Air Force. Only one additional contract flight was used and that was for the expeditious movement of the final increment of personnel following the 26 Nov 91 closing ceremony.

Flooding along the road between Clark and Subic caused delays in moving freight to the port and threatened to stop movement altogether--necessitating an alternative. Other routes to Subic were no less susceptible to closure than the primary route. Therefore, we coordinated with NSD, MTMC, PACAF, APL, SeaLand and TMO, Manila to establish procedures and rates to move freight out of the port of Manila expeditiously.

HO PACAF/LGTV assessed the vehicle fleet and found that the fleet has sustained both cosmetic and internal damage. We sent a list, by regulation number through PACAF to Airstaff, detailing those vehicles we proposed to leave in place. The list was approved, with minor deletions. All vehicles which HHQ originally said to ship were shipped. Likewise, we transferred several excess GOV's to other on-island DOD and Department of State organizations.

WEATHER

All required suspenses were met on or before the suspense dates. Of $877,863 of equipment (per CNCRL), $101,377 was left in place (11.5%), mainly weather communications equipment that was too costly to remove. The weather radar (50K) shipped out 15 Nov 91 and the Low Resolution Weather Satellite Monitor (Kenwood Looper -60K) on 22 Nov 91. This equipment was kept in place as long as possible to provide reliable data in case of a late November typhoon.

All individuals PCS'ed at or prior to their scheduled departure date. While we encountered no real "showstoppers", our biggest challenges were:

1) coordinating places to send equipment;
2) determining what needed to be done when;
3) managing drawdown of equipment/personnel and associated required actions.

Part of our solution was to hold biweekly staff meetings that determined where we were, what needed to be done and to delegate tasks.

13TH AIR FORCE

OPERATIONS: Following the eruption of Mt. Pinatubo the first elements of the DCS Operations staff returned to Clark on 26 June 91. Subsequently, building 2125 was prepared to serve as an emergency safe haven for the entire 13AF staff and MREs and water were stocked for a contingency. Additionally, Operations COMSEC was turned in to the 1961 CG and accounts closed. Operation's classified documents were segregated and prepared for destruction. Although C2 assets had moved to Kadena the operations staff continued to coordinate OSA support for the 13AF/CC and documented future C-21 OSA requirements for 13AF support at Guam. Additionally, the transfer of USAF allocated airspace to the US Navy was coordinated through Villamor Air Base. 13AF operations also managed the overall transfer of all 13AF headquarters equipment and furniture in 13 sea/land containers to Andersen AFB, Guam.

PLANS AND PROGRAMS: Following the closure announcement, Plans and Programs staff initiated a review of it's classified holdings which consisted primarily of plans documents (eight safes worth of material). Subsequently5 a systematic process was begun for the destruction of plans which were either not critical to 13AF's mission or could be reordered from the plan OPR. The remaining classified documents were then mailed or sent to Guam via the Defense Courier Service as appropriate by 22 Nov 91. During this period, XP continued to support 13AF/CC regional visits with trip books and U.S.-Phil Politico-Military and Mutual Defense Board activities. In addition, 13AF/XP coordinated the procedures for documenting the transfer of Wallace AS, Camp O'Donnell and Clark AB real property and excess defense articles to the PAF with SAF/IARP and HQ USAF/XOXXP.

CONCLUSION

In closing, these were the mission objectives that guided our actions: safety . . . dignity . . . communication . . . and the preservation of the heritage of the oldest continuously active fighter wing in our Air Force.

On the 26th of November, when the American Flag was lowered for the last time and presented to Ambassador Wisner, it was done in keeping with those objectives and that tradition. It also marked the end of an U.S. military presence here that started more than 90 years ago.

BRUCE M. FREEMAN
Colonel, USAF
Commander

Appendix 2 VOLCANO EVACUATION PLAN PAMPHLET

AS OF 1700 L, 7 JUNE 1991

INTRODUCTION

Mt. Pinatubo has not erupted for over 600 years. However, recent events suggest that this could change. As a result, U.S. and Philippine experts are closely monitoring the volcano, 24 hours a day. Also, the Clark AB command authorities are providing this pamphlet for your use to plan accordingly.

Keep in mind that even with our best efforts, several hours may be all we have before the volcano erupts. This might happen at any time day or night. The bottom line? To prepare in advance to ensure your safety and the safety of your family.

This pamphlet gives you the best information based on the circumstances at the time it was printed. Your unit is keeping an up-to-date plan with checklists and things that need to be done. Stay in contact with your First Sergeant or Commander for information.

A final note. This pamphlet was based on a "worst case scenario" which may never happen. Your unit will keep you informed as to how things are going on the mountain and what steps you should be taking.

Again our immediate concern is your safety.

CONTENTS

1. Where you are going.

If we are required to evacuate, all non-essential personnel, dependents, and pets will go to Subic Naval Base. However, with your commander/First Sgt approval you can go elsewhere if you provide them with your address and phone number. The Navy has plans to house and feed everyone from Clark on an emergency basis.

Our immediate concern will be to get everyone safely to Subic and out of harms way. Once we are re-established at Subic, senior leadership will evaluate the situation and determine how long we will stay and whether we return to Clark or go on to some other destination.

For off-base personnel, you are not in safe areas listen to FEN for definitive guidance.

Upon arrival at Subic, you will be directed where to go and what to do. You will be met by your organizational representative at the Subic Sampiguita Club (bldg 418) who will provide you further assistance.

2. What you should plan to take with you.

Everyone should start right now to build a "Bug-Out" kit containing what you will need. This should be kept ready to go at all times. The list of items below gives you things to consider which you can tailor to fit your particular circumstances.

Clark residents waited on the flightline for the signal to evacuate.

ITEMS YOU NEED TO TAKE

1. Cash for each family member (both dollars and-pesos)
2. I.D. cards
3. Passports/Visas
4. Ration cards (CEX)
5. Immunization records
6. Birth certificates
7. Marriage certificate
8. Toiletries
9. Three sets of clothing for each family member
10. Pillow and blanket or sleeping bag for each individual.
11. Vehicle registration and car insurance policies
12. Checkbook/Bankbook/Credit Cards
13. Personal insurance policies
14. Inventory of all household goods
15. Will (s)
16. Personal medications
17. Infant care items (diapers, toys, formula, portable crib, stroller, etc)
18. Water supply for 1 day
19. Food/snacks for 1 day
20. Food for pets (3 days) and leash
21. POVs should have 1/2 to 3/4 tank of fuel
22. Flashlights with spare batteries
23. Portable radio with spare batteries
24. Candles/matches
25. Bug repellent
26. Sun screen hats, etc.
27. First aid kit, knife, rope, tool kit
28. Toilet paper

NOTE UPON LEAVING QUARTERS:

1. Turn off all airconditioners/central air or other ventilation devices
2. Turn off electric ranges, washers and dryers. Keep refrigerator and freezers on.
3. Close all windows.
4. Leave some interior and all exterior lights on.
5. Lock all external doors.

3. When and how you are going to get off-base.

a. The following guidelines will help assure an orderly flow from the housing areas.

b. You will be notified to leave Clark AB by:

(1) Unit recall roster

(2) Announcements on FEN

(3) Base siren sounding a steady tone for five minutes. If this occurs, tune in to FEN radio/TV for further information.

(4) The Security Police using their vehicle PA system in the housing areas

(5) Giant Voice (flightline speaker system) broadcasts

c. Once you are notified, everyone except mission essential personnel will prepare to evacuate to Subic Naval Station.

d. Pick up your "Bug Out" kit and use your POV to go to the flightline following the routes on the attached maps. Attachment 1 is the Base Evacuation Route and Attachment 2 is the Route to Subic.

(1) Personnel with automobiles should depart immediately to the flightline area. Take everyone in your household including your domestic help and pets. Use the buddy system. Check the house to the right and left of you for occupants. Do not forget anyone!

(2) Those without automobiles arrange to catch a ride with a neighbor or squadron member. Military family housing occupants should pick up any members they see needing a ride. Be friendly!!

(3) Note that cabs may or may not be running.

(4) If you have been left in the housing area without transportation, call the Sub Motor Pool 393-3341/42 or the Security Police 393-3484.

(5) Should evacuation of off-base people become necessary, military, contractor, DoDDS, and US civilian personnel, and their dependents who need transportation should report to the Quad Agency Patrol (QAP) area at the Main Gate. Show your ID card when you arrive and the QAP will make transportation arrangements.

e. TRAFFIC FLOW (See Maps)

(1) To prevent major traffic backups at the various gates, families from base housing will be directed to the flightline on three major routes. Security Police will be at all major intersections to direct traffic smoothly.

(a) Mitchell Highway (Sector I): Houses in the hill area west of where Mitchell intersects Anderson and those houses south of Mitchell highway as it passes through the main base will be directed onto the flightline at the 1st Test/South ramp entry control point and further directed off base

(b) Dyess Highway (Sector II). Families in the hill area east of Mitchell highway should proceed down Anderson or Dyess then proceed down Dyess Ave where they will be directed to the flightline and then off base.

(c) Bong Highway (Sector III). Families in the main base area (Sector III) will be funneled to Bong highway and then directed onto the flightline and off base.

(2) The primary exit will be through the Dau gate and then onto the Northern Luzon Expressway. If alternate exits/or routes are used, personnel will be directed by the Security Police.

(3) Once you enter the expressway, head south toward Manila. Exit at San Fernando. Turn right and continue straight ahead through the traffic light. Your next major turn is a Dinalupihan where you again keep right and head toward the mountains. There will be joint US/RP patrols at all intersections to provide assistance. There is no need to be in a hurry- -make this a safe trip. There will be patrols along the route to provide assistance if you break down.

4. What to do when you get to Subic.

a. When you arrive at Subic, follow the instructions 0f the Navy personnel for quarters assignment and of AF PERSCO personnel located in Bldg 415 (The Subic Sampiguita Club) for inprocessing. Advance teams from your military unit will be set up to assist you. Provide your name, organization, SSAN, names of family members with you, and of any who may be missing.

b. After you inprocess, Subic will provide transportation to your Billeting location.

c. Remember, take care of your families. Stay tuned to the various FEN AM, FM, and TV channels for more information. You may also get official AF news from the Manila media.

5. What you will do when the emergency is over.

a. We will keep you advised, so stay tuned to FEN.

b. There are several things that could occur.

(1) Worst case, Clark Air Base would not be habitable. Should this happen, you would stay at Subic until the AF could make arrangements for you to leave. Some of you would be sent to the US. Others could volunteer for a continuous overseas tour (COT) and, if there is a need for your specialty, be sent to another overseas base. Since volcano damage would be considered total, you would need to make your personal property claims through the legal office. In this case, legal personnel would be available to assist you. We will let you know over FEN how to contact legal to make your claims.

(2) Best case, we could return to Clark. Upon return, carefully inventory your belongings and check for damage. You have the right to make a claim for damaged or missing items. Leadership requests you report all facilities damage to civil engineering personnel, but otherwise begin cleaning up your immediate area. We'll provide you information on what to do first to minimize possible damage or injury. Again, listen to FEN for further guidance.

6. Volcano Characteristics

a. Ash. Fine material blasted aloft by a vertical explosion from the volcano that could reach an altitude of 12 or more miles. As the ash cloud spreads it could be dense enough to screen out nearly all sunlight. While ash will be deposited over an enormous area, the heaviest deposits will be on the downwind side of the volcano. Ash is a hazard to driving and all things of a mechanical or electronic nature, including telephones, computers, and aircraft.

b. Pyroclastic Flows. Are hot, often incandescent mixtures of volcanic fragments and gases that sweep along close to the ground. Depending on the volume of material, proportion of solids to gas, temperature, and slope gradient, they can travel in excess of 100K/H. They are extremely destructive and deadly because of their high temperature, which ranges from 450 to 1000 degrees C.

c. Mud Flows. Are mobile mixtures of volcanic debris and water. Where water is available to erode and transport the loose deposits of debris, that are very destructive, could begin within 50 minutes after the start of an eruption, and will follow existing river channels.

d. Lava Flows. Are molten rock that emerge from volcanoes. They are very hot, but move very slowly, and are not a factor at Clark.

e. Projectiles. Are airborne products of a volcanic eruption that could be hurled a distance of about 10K from the volcano. We plan to move you well outside that area and provide you with an additional margin of safety.

7. Questions and Answers

1. How long will the eruption last, assuming it does erupt?

It could be for a day or from two to three months or more.

2. Can the volcano emit water and what happens if it does?

Not this volcano. However, rainfall could cause very dangerous mud flows and/or flash floods down local valleys.

3. If it erupts, how far could it hurl projectiles?

While it is impossible to predict accurately, pieces of rock weighing several tons could be blasted up to ten Kilometer.

4. Are fissures in the earth possible?

Yes. They could open and close very quickly, but there is no danger of this at Clark. The greatest danger is on Mt Pinatubo, where we shouldn't be under any circumstance.

5. What is the nature of a lava flow?

As you know, lava is liquid rock that has all the characteristics of fluids. As it flows down hill, it will follow the sloping terrain. Occasionally, a river of lava will split and split again, only to rejoin later trapping all those unfortunate enough not to have gotten away. Those trapped will probably die from either the heat or asphyxiation. However, it is very slow moving and we don't expect much of this from this volcano.

6. Could this volcano explode like Mt St Helens? If so what would it be like?

Volcanoes are unpredictable. We think it could explode because all volcanoes have that possibility given the right circumstances. If it does, the noise of the explosion would probably be heard throughout Luzon, and possibly in Viet Nam. The blast could be up to or in excess of 100 miles per hour and carry temperatures from 450 to 1000 degrees (centigrade)

7. Is there any danger from the sulfurous air?

It depends. In small quantities, there is not much danger. The human nose will detect hydrogen sulfide in trace quantities, but there are no medical effects. Sulfur dioxide is a respiratory irritant. Both chemicals are being monitored in the air and will be posted similar to smog alerts in LA. Alerts will be announced on FEN.

8. How many active volcanoes are there in the world?

There are about 500 active ones at this time. Several thousand are extinct, while others located under the sea, both active and extinct, are unknown.

9. Isn't this like the volcano in Hawaii? Why should I be concerned?

While all volcanoes are related in their potential for destruction, some are reasonably predictable. The volcanoes in Hawaii have been active for some time. We know what to expect from them. Others, like Pelee on the Caribbean island of Martinique, exploded violently killing over 30,000 in one day. Two years later, just as people were beginning to return, it erupted violently again, this time killing over 2,000. Another volcano, Krakatoa, on an island between Java and Sumatra exploded violently and the blast was heard over a thousand miles away. The volcano destroyed itself and the island. Mt Pinatubo is relatively unpredictable, since it hasn't erupted in over 500 years. The magma from Pinatubo is a thick, explosive type; while that from Hawaii is a fluid, relatively non-explosive type. Not knowing exactly what type eruption to expect, we must proceed cautiously, advise everyone to be concerned, and to plan for the worst. If the eruption is less than it could be, we will all be relieved.

10. How long would we have to stay away before we are allowed to return to Clark? What if we can't come back?

We will have to determine this based upon USGS analysis and senior management judgement. If the base is habitable, we'll let you return as soon as it is safe for you to do so.

If the base is destroyed, or the decision is made that we should not return for reasons that we have not yet considered, we will either transport you to another US military installation, such as Hickam for reassignment, or reassign you from Subic.

11. What will become of all our things? Can we go back and get them'?

If the base is habitable, you'll be allowed to return to your quarters. When you return, carefully inventory your items and report damage or loss to the legal office.

If the base is not habitable, you will not be allowed to return. You should then file a claim with the legal office.

12. If we evacuate Clark, can we take our domestics and our pets?

The answer is yes. However, facilities at Subic are limited and dedicated to US military, contractors, DODDS, US employees, and dependents. While it is not feasible to take domestics to Subic, it is imperative that we not leave them behind on-base because to do so could mean their death. They should be removed from the housing area and, once on a safe part of base, allowed to go their way and join their friends and family.

13. Do you anticipate a traffic problem between here and Subic or Manila?

Absolutely. The l.5 to 2-hour ride to Subic could be greatly extended. Based on the traffic we believe will be on the road and assuming the road stays bi-directional, you could expect a 6 hour plus trip. However, we are working with Civil Defense authorities to turn the highways into "away" traffic only. That would make a two lane single direction highway from here to Subic and as much as a 5 or 8 lane highway into Manila, and greatly enhance traffic flow. But you should not count on single direction traffic. Make sure you have enough gasoline.

14. How can you ensure all people are removed from Clark?

No system will be error proof. However, if we use the buddy system, by that I mean check the houses to your right and left and physically ensure no one is left behind, we can be reasonably sure we get everyone out. This will work day or night. The same procedure should be used in the dorms.

Remember, we expect to have sufficient notification, so don't panic.

-signed-

JOHN E. MURPHY, Colonel, USAF

Commander

Appendix 3 MISSION ESSENTIAL LIST AS OF JULY 20, 1991

ABCEDE, RUDOLFO P C
ABELLANA, JARVIS A
ABINSAY, JERRY P
ABUEG, BLESILDA C
ABUYEN, URBITO A
ACKER, BOBBY D
ADAMS, C D
ADAMS, NEWELL F
ADAMS, RONALD L
ADAMS, TRACY L
ADKINS, RICKY J
AGNEW, SEAN P
AGUILAR, RICHARD L
AGUSTIN, LEONIDES B
AINARDI, DEAN M
AINSLIE, RONALD J
ALBIN WAYNE F
ALBRIGHT, ANDREW C
ALBRIGHT, JACOUELINE
ALDANA, B Q
ALEXANDER, M
ALFONSO, EDDIE S
ALLEN, DENNIE M
ALLEN, GARY D
ALLEN, HOLLY J
ALLEN, KERMIT D JR
ALLEN, PHYLLIS E
ALLEN, RICHARD D
ALLEN, ROBERT W
ALLGOOD, KENNETH W
ALLOWAY, VIRDA L
ALMACEN, CHEN G
ALMACEN, DWIGHT
ALMACEN, TERESITA M
ALMARAZ, REYNALDO I
ALTENBERND, R L
ALVAREZ, JOSE A
ALVAREZ, ROGER E
ALVES, RAUL J
ALVEY, RONALD E
AMASOL, TITO P JR
AMES, RICHARD S II
AMOS, JARVIS V
AMPOYO, GEOFFREY A
AMURO, ASAICHI
ANDARIN, ROSENDO I
ANDEL, JAMES
ANDEREGG, C R
ANDERSEN, LOIDA C
ANDERSON, C
ANDERSON, DANIEL L
ANDERSON, D W
ANDERSON, ERIK C
ANDERSON, JAMES L J
ANDERSON, JOHNNY R
ANDERSON, JUNIOR L
ANDERSON, M W
ANDERSON, TIMOTHY A
ANDERSON, TOBIAS M
ANDERSON, WILLIAM C
ANDREWS, GARY C
ANDRZEJEWSKI, BRIAN K
ANGYAL, NOLA S
ANTHONY, JAMES E
ANTHONY, LEWIS
ANTOINE, DEBBIE J
APODACA, ISIDRO
APPLEWHITE, JESSE E IV
ARCADI, TIMOTHY W
ARCHBOLD, BRET E
ARD, TIMOTHY D
ARELLANO, HERMINIO M
ARENAS, MARCOS A L M
ARMFIELD, HARVEY L
ARMSTRONG, WILLIAM F
ARRIETA, JAIME
ARROJADO, GLENNA M
ARROYO, PIO S III
ARTHUR, LOUIS W
ARVIN, DANIEL
ASH, GUY P
ASHBAUGH, DONALD E JR
ASHLEY, SCOTT C
ASHMORE, GARY A
ASTLE, LAWRENCE D JR
ATCHLEY, CHARLES Y
ATIENZA, PROSPERO JR
AURELIO, HONORIO F JR

Ash Warriors received briefings on what to expect.

AUSTIN, TYPONE JR
AUZENNE, JAMES H
AVENA, DAVID M
AVRITT, CHARLES E
AYRES, RONALD L
AYZE, THOMAS M
BABER, DANIEL S
BABICS, RONALD J
BABIN, JAMES M
BACHER, GREGORY D
BADER, ROBERT E JR
BAGBY, LARRY P
BAILEY, CHRISTOPHER L
BAILEY, EDWARD A
BAILEY, JOHN E JR
BAINBRIDGE, JAMES W
BAKER, JAMES H
BAKER, JOHN E
BAKER, KARLTON L
BAKER, MARK C
BAKER, MARKE A
BALDWIN, DAVID C
BALDWIN, HOWARD D
BALEY, JOHN F
BALISI, EFREN C
BALL, GARY A
BALLENTINE, STEPHEN A
BALLINGER, JEFFREY L
BALLOU, ELEANOR F
BALLUCANAG, RICHARD C
BANDELOW, JAMES J
BANEZ, ELPIDID B JR
BANKS, JOSEPH K
BANKSTON, WADE D
BAON, RAUL
BARBA, RAY A
BARCEGA, JESSE E
BARDELL, THOMAS A JR
BARE, WILLIAM H
BARGERY, CHRIS
BARKER, STEVEN G
BARKLEY, JON P
BARKOW, LAWRENCE J II
BARNES, DARRYL D
BARNES, DAVID R
BARNES, JOE A JR
BARNES, WILLIAM E JR
BARNETT, JAMES L
BARNETT, MATTHEW J
BARNETTE, DANIEL J
BARNHART, LISA D
BARRERA, JOSE
BARRETT, LISA A
BARRETTE, RANDALL L
BARRETTO, PHILLIP
BARRON, RICHARD M JR
BARSY, DALE A
BARTA, ADRIAN C
BARTLEMUS, STEVEN P
BARTLETT, JONATHAN B
BARTLETT, LEE D P

BASS, CLEON E
BASS, TIMOTHY L
BATALON, RODERICK
BATHURST, THOMAS F
BATINO, LAUDAN J
BATOR, KRIS
BATTLE, DUSHUN 0
BAUMGARTEN, DOUGLAS
BAUTISTA, BRADLEY E
BAXLEY, JAMLS A
BAYLOR, DIETRICK S
BAYNARD, CHARLES E
BAZAR, LOUIS A JR
BEALE, WILLIAM
BEAR, CHARLES F JR
BEASLEY, DEWEY L
BEATY, KEVIN S
BEATY MARK A
BEAUCHENE, WILLIAM R
BECK, JAMES D
BECK, MICHAEL J
BECK, TIMOTHY A
BECKER, SHELLY L
BEDARD, RYAN E
BEECH, FLOYD R
BEERS, BRYON M
BEHARRY, RONNIE T
BEHN, WILLIAM G
BELISLE, GARY R
BELL, DURWOOD A
BELL, JERI A
BEN, REYNANTE C
BENETSKY, STEPHANIE
BENNETT, CHRIS E
BENNETT, DAVID W
BENNETT, HOWARD A JR
BENNETT, MARK L
BENSON, MARY H
BENSON, PHILLIP M
BERCK, TIMOTHY D
BERGMAN, BRADLEY A
BERGREN, RANDEL L
BERRY, DARRELL E
BERRY, KELLY D
BETTIS, TEDDY L
BETZ, KENNETH R
BEXELL, SCOTT D
BIPES, JANIS L
BIPES, MICHAEL D
BIRD, JAMES W
BIRD, RALPH M
BISHOP, BRADLEY M
BISHOP, KENNETH A
BIVENS, CYNTHIA A
BIVENS, GLENN E JR
BIVINGS, LEEVAN JR
BLACHOWSKI, CHAD G
BLACKMON, JOHN M
BLACKMON, ROBERT F
BLAIR, BRIAN S
BLAIR, KENNETH D

BLAKNEY, CHARLES V
BLANCHETTE, JESSICA A
BLANKS, LAWRENCE T
BLASQUEZ, WILBERTO S
BLATUS, ANDREW S
BLESSINGER, MICHAEL J
BLOOM, RICHARD W
BLYSTAD, THOMAS J
BOBBITT, JAMES C
BOCOOK, CLARENCE R JR
BODISON, JAMES L
BODON, FRANK R JR
BOGER, WAYNE L
BOGUCKI, STANLEY J
BOLICK, ROBERT L
BOLIN, DONOVAN A
BOLTON, SCOTTIE R
BOLUS, ANNIE C
BOLUS, ROBERT
BOND, DOUGLAS W
BONIOR JOHN E
BOOKER, DAVID B
BORTZ, CHALMER R JR
BOSTICK, STEVEN F
BOSTON, LARRY D
BOULDREY, COREY A
BOULWARE, STEPHONE F
BOURGEOIS, DEBORAH A
BOWEN, BRIAN S
BOWEN, JAMES E
BOWEN, JEFFREY T
BOWMAN DIANE M
BOYD, JOHN C
BOYD, PATRICK W
BOYER, MARK D
BOYLAN, KIRK F
BOZZELLI, ALFRED C
BRACKETT, JAMES
BRAEMER, RAYMOND W JR
BRANDT MICHEL C
BRANTING, JEFFREY G
BRANTON, JEFF
BRANTON, LAURA
BRAUD, RICHARD K
BRAZELL, ERIC J
BRAZZLE, TRACY C
BRENHOLT, MATTHEW D
BREUER, MATTHEW S
BREWER, KENNETH W
BREWER, TIMOTHY A H
BRIDGES, CHARLES R
BRIDGES, REGINALD D
BRIDGEWATER, RUSSELL L
BRIGGS, BOBBY S
BRIGGS, KENNETH M
BRISLEY, KENNETH E
BRISLEY, KEVIN E
BRISSETTE, ROGER M
BRITT, THOMAS H
BRITTON, JOSEPH A
BRITTON, MARIO D

BROECKER, JON E
BROOKS, BRIAN J
BROOKS, DAVID O
BROOKS, FORD A
BROOKS, RICHARD M
BROOME, KENNETH A
BROTHERS, EUGENE A
BROWDER, JAMES F
BROWER, BRUCE
BROWN, CLARENCE G
BROWN, DAVID D
BROWN, JAMES E
BROWN, LARRY A
BROWN, MICHAEL T
BROWN, MORGAN D
BROWN, ROBERT J JR
BROWN, STEVEN E
BROWN, TIMOTHY L
BROWN, WILLIAM E
BRUCE, DESI W
BRUCE, RICHARD J
BRUCE, TERENCE C
BRUDER, RICHARD P
BRUKWICKI, STEVEN E
BRUNO, PAUL J JR
BRUTON, GLEN W
BRYANT, HEATHER J
BRYANT, HENRY A
BRYANT, MIKE A
BRYNER, STEVEN J
BRYSON, CHRISTOPHER L
BUCANEG, RUFINO JR
BUCKMASTER, ROY C JR
BUENSUCESO, RENATO A
BUFFINGTON, JAY R
BUIKUS, ROBERT J
BUMGARNER, RALPH I
BURGESS, LAWRENCE J
BURKE, RICHARD D
BURKE, WILLIAM P JR
BURKETT, STEVEN D
BURNETT, LAPARIS
BURNETTE, JAMES S
BURR, JOSEPH W
BURROWS, ALFRED D
BURT, ALAN
BUSH, ANDRE C
BUSH, MARK J
BUSH, TIMOTHY E
BUSHEY, DEAN E
BUSSEY, JONATHAN
BUTLER, DARRIN C
BUTLER, GREGORY R
BUTLER, PHILLIP C
BUTLER, SCOTT D
BUTLER, WAYNE L
BUXTON, MARTHA J
BUXTON, ROBERT G
BYERS, RODNEY L
BYFORD, RICHARD L
BYRD, REX A

CABALLERO, ARSENIO V
CABANA, ISABELITO M
CABANERO, CHARLES E
CADALBERT, MARK A
CADLE, CLAYTON D
CALAHAN, RANDO B
CALALO, DANNIE
CALBAY, ALDY M
CALES, ROBIN L
CALEY, CHARLES W JR
CALHOUN, JOHN G
CALHOUN, MICHAEL R
CALHOUN, ROBERT K
CALLAIS, CHRISTOPHER M
CALLAN, KEVIN T
CALLOWAY, RODNEY
CALVERT, JAMES L
CAMACHO, ENRIQUE
CAMERON, RICHARD N
CAMPBELL, BRUCE I
CAMPBELL, LARRY S
CAMPBELL, ROBERT A
CAMPBELL, ROBERT R JR
CAMPBELL, WELDON R
CAMPLIN, JON
CAMPOS, HELIODORO J
CANAGUIER, ROBERT M
CANALES, JESUS M JR
CANDA, RAMON D
CANLAS, JORGE N
CANNELLA, ARTURO A
CANNON, DOUGLAS
CANNON, JAY R
CANTRELL, DONALD R
CANTU, JUAN G
CAPPS, ALBERT T
CAPUNO, LEE M
CARBNO, MICHAEL J
CARDENAS, ISRAEL
CARDENAS, LOTHAR G
CAREY, DEANA M
CARGES, MICHAEL R
CARINO, ART
CARLSON, THOR
CARMACK, ROY J
CARMICHAEL, CORY D
CAROTHERS, SAMUEL C
CARPENTER, SHERRI A
CARPENTER, TERRY G
CARREON, EDGARDO D
CARRINGTON, ROYLE
CARROLL, DOUGLAS F
CARSLEY, JOE
CARSON, EDWARD J
CARSWELL, KEVIN A
CARTER, ANGELA L
CARTER, PATSY S
CARTER, ROBERT D
CARTWRIGHT, TYRONE
CARUTHERS, SEAN S
CASDEN, MARC F

CASEM, MANUEL I
CASEY, MARGARET H
CASTILLO, ADONIS B
CASTILLO, DANTE A, CHRISTO-
PHER J
CHEEK, JAMES H III
CHERRY, JOHN E
CHERRY, MELANIE SUE
CHERRY, SCOTT A
CHISSEM, DENICE A
CHRISTENSEN, JEFFERY A
CHRISTIE, WALTER C JR
CLARK, HARRY
CLARK, HENRY J
CLARK, JACK M JR
CLARK, RICHARD C
CLARK, RGNALD E
CLARK, TOMMY JR
CLAYBROOKS, CLIFTON JR
CLAYTON, CORNELIUS F III
CLEEK, DANIEL W
CLEGG, JOHN D
CLEMENSON, MICHAEL J
COLE, JOHN F
COLE, JOSEPH H JR
COLEMAN, RUFUS W JR
COLEMAN, TRAVIS
COLHOUER, RAYMOND T
COLLINS, CALVIN C
COLLINS, KEVIN A
COLLINS, TERRY D
COLSON, DAVID M
COMEAUX, ROBERT M
COMER, JEFFERY L
COMMITTEE, DENNIS M
COMPTON, DENNIS L
COMPION, VERNON
CONCORD, ALBERT L
CONE, ROBERT J
CONGDON, WILLIAM A
CONGER, RICHARD A
CONNELL, POLLY A
CONNERS, STEVEN L
CONNOLLY, ANNETTE
CONSUEGRA, EDWARD S
CONTRERAS, DAVID
CONVERSE, MICHAEL S
COOK, MICHEAL R
COOK, PAUL
COOPER, DAVID J SR
COOPER, JOHN R
COOPER, PAMELA D
COPERTINO, DANIEL
COPPEDGE, DARRELL W
CORBETT, CLARENCE A
COREY, JAMIS L
CORIELL, DENNIS
CORLEE, KENNETH L
CORLISS, WALTER F II
CORNELL, MATTHEW J
CORNS, EDDIE R

CORNWELL, BRUCE A
CORPUZ, DANILO M
CORWIN, ARTHUR J
COSTON, CAREY L
COTTER, ALBERT L
COTTLE, JAMES E
COTTMAN, BARRY
COTTON, RAYMOND E
COUGHLIN, CORY A
COUGHLIN, RICHARD
COUNTY, WILLIE JR
COUPLAND, JAMES P
COURTRIGHT, PAUL F
COUTEE, JOE H JR
COWART, ROBERT J JR
COX, MELVIN
CRABTREE, DAVID E
CRAMER, KENNETH A
CRAVEN, ROBERT P
CREEL, MICHAEL H
CRESPO, WIEFREDO
CRESS, GARY W
CROCKETT, KEVIN M
CROMARTIE, CHARLIE C
CROOK, LINCOLN S
CROOMS, STEPHEN C
CROSS, KEITH A
CROWE, MARVIN E II
CROZIER, DONALD E JR
CRUMBO, KEVIN M
CRUZ, ROBERT C
CUEVA, LORETO N
CULBERTSON, DANIEL R
CULBRETH, ROBERT S
CUMMINGS, JOHN C
CUMMINGS, TOOD L
CUNNINGHAM, BRIAN K
CUNNINGHAM, JAMES W JR
CUNNINGHAM, LESTER
CURRIE, SHAWN D
CURRY, BRETT A
CURRY, WILLIAM P JR
CURTIS, LESLIE D
CUTSHAW, JERRY D
CZWALINA, PAUL P
DADE, FRANK E
DAITCH, SHELDON
DALE, JIMMIE D
DALE, STEPHAN H
DALTON, AMY E
DALTON, JUDY D
DALTON, SCOTT F
DALY, THOMAS J
DAMBITIS, TAMMY D
DAMEWOOD, GERALD L
DANIELS, WILLIAM J
DANNA, ANGELA M
DANZ, EDWARD J
DARBY, KENNETH R
DARKEY, ROBERT W II
DARR, DAVID S
DASSLER, WILLIAM H
DATAR, EDGARDO A JR
DAUFEN, WILLIAM K
DAVENPORT, KEVIN L
DAVILA JERRY R
DAVIS, CHRISTOPHER R
DAVIS, DAVID M
DAVIS, DAVID T
DAVIS, DWRIGHT J
DAVIS, GERALD A JR
DAVIS, IRVING A
DAVIS, JAMES M
DAVIS, JEFFREY W
DAVIS, JERRY J
DAVIS, JOHNNY L
DAVIS, KEITH D JR
DAVIS, KEVIN J
DAVIS, KEVIN L
DAVIS, RAYMOND S
DAVIS, REX W
DAVIS, RICHARD W
DAVIS, ROBERT E
DAVIS, STEVAN E
DAVIS, VINCENT A
DAVIS, WILLIAM D
DAVITT, DENNIS K JR
DAWSON, JOHN E
DAWSON, MARK H
DAWSON, TERRY L
DEALOIA, JOSEPH
DEAN, MARVIN R
DEANS, GREG
DEARING, FREDERICK L
DEBELLIS, PATRICK T
DELACRUZ, J C
DEEACRUZ, WILFREDO G
DELAROSA, ALLER B
DELAVEGA, DOMINIC E
DELL, KENNETH C
DELMUNDO, ISAURO B
DELUNA, EDWARD E
DEMASS, MICHAEL J
DEMONT, ENRIQUE
DEMONTE, LOUIS R
DEMPSEY, DANNY L
DEMPSEY, JEEFRY H
DENIZ, FRANCIS R L
DENNIS, WILLIE H
DENNISTON, ROBERT B
DENTON, ANTHONY E
DERLEIN, DANIEL J
DESCH, ANTHONY J
DESMOND, JOHN J
DESROCHERS, KEVIN D
DEVRIES, DAVID M
DEWEY, JAMES H
DEYOUNG, JEFFREY L
DIAL, DAVID
DICKERSON, BRIAN K
DICKERSON, LAURENCE F
DIEDRICH, PAUL F
DIEFENBACHER, STFVEN P
DIETRICH, CHRISTOPHER
DIETZ, MARK A
DIFIORI, THOMAS C
DIGANGI, WILLIAM A
DILLON, DALE M
DIRCKX, VINCENT J JR
DIRITDO, STEVE W
DISANTO, STEPHANIE
DISSELHORST, SCOTT A
DITTMER, GLENN R
DIXON, BRUCE W
DIZON, CHRISTOPHER M
DOBBINS, WILLIAM
DOBOGAI, DARRYL A
DOCTOR, DEREK L
DODD, JOSE A
DODD, KENNETH R
DOE, GERALD M
DOLPHIN, ARLENE
DOMINSKI, WILLIAM P
DONAHUE, CHRISTINE L
DONNELL, JOHNNY D
DORAM, RAPHAEL O
DORAN, ALBERT J
DORAN, MARTIN J
DORSEY, GARY B
DOUGHERTY, JOSEPH T
DOUGHTY, GORDON K JR
DOUGLAS, CHARLEE
DOUGLAS, CHARLES K
DOUGLAS, KENNETH E
DOUGLAS, STEPHEN W III
DOWNES, JOSEPH L
DOWNS, DENNIS L JR
DOYEA, THOMAS P
DOYLE, BRIAN P
DRAEVING, JAMES E
DRAKE, GARRY D
DRAKE, HAROLD
DRAKES, LUIS A
DREW, MICHAEL G
DUARTE, JOHN S
DUBASIK, DEREK M
DUBOIS, ERIK B
DUDA, JAMES M
DUDLEY, JO A
DUERNBERGER, RODNEY L
DUFF, CLARENCE E
DUFF, ROGER L
DUFFY, BRIAN J
DUFFY, STEPHEN L
DUHART, GENTRIS Y
DUKE, CHRISTOPHER A
DUKE, RAYMOND
DUKE, ROBERT J
DUMDUMAYA, RICKY M B
DUMLAO, BASILIO F
DUMLAO, EDWIN S
DUNGAO, BASILIO R
DUNSMORE, TIMOTHY A

DURHAM, DALE J
DUTCHER, GERALD B
DWIGGINS, DAVID W
DYE, SAMUEL R JR
EAKLE, BENJAMINE H JR
EARNEST, JOHN E
EBINGER, MARC A
ECKLES, WILLIAM J
ECKERT, RONALD G
EDDER, JERRY W
EDGAR, SCOTT M
EDMONDS, MAURICE JR
EGGERT, STEPHEN R
EIDE, ERIC J
EIKER, DAVID M
EISENHARDT, ROBERT A
ELKINS, JODY R
ELLIOTT, JASON D
ELLIOTT, ROGER A JR
ELLIOTT, TIMOTHY E
ELLIS, CARL H
ELLIS, DONALD E
ELLIS, GARRY W
ELLIS, REGINALD J
ELLISON, WILLIAM J II
ELLSWORTH, KENNETH G
ELVEY, W M
ENDERS, GARY A
ENGEL, PETER
ENGELHARDT, KENNETH E
ENGELKES, ARDEN B
ENGLISH, LILBURN JR
EPPS, MAURICE S
ERGENBRIGHT, BRENDA J
ERGENBRIGHT, WILLIAM B JR
ERIKSON, ROGER A
ERKENEFF, TERRY L
ERNST, DOUGLAS N
ESPEJO, JESUS C
ESPEJO, ROMAN H
ESPIRITU, ANGEL P
ESTABROOK, BRUCE A
ESTES, PAUL S JR
ESTIGOY, EMMANUEL L
ESTIRA, MICHAEL O
ESTRADA LUCINDAM
ETIENNE, WILTON A
ETTENFELDER, J
EVANGELISTA, DENNIS A
EVANS, ESTLE R
EVANS, KEVIN J
EVANS, LEONOLD R
EVANS, MICHAEL N
EVANS, MICHAEL W
EVANS, ROBERT E
EVERINGHAM, RICHARD O
EVERSELY, SAMUEL S
EVLITT, WILLIAM
EYMAN, MARK E
FABER, GREGORY L
FAILS, DEBORAH
FAILS, GEORGE L
FAIRFAX, DAVIS J
FALES, VICKY L
FALLEN, KEVIN E
FARIAS, ROY S
FARMER, JEFFERY C
FARMER, ROY F
FARR, STEVE M
FARRELL, GREGORY L
FARRELL, RICHARD J
FARRINGTON, PAUL A
FE LDBAUER, ROBERT R
FELDT, WILLIAM M
FELIX, JAIME D
FELLMAN, BRYAN R
FELLOWS, RONALD G
FENWICK, ROBERT W
FERNS, TODD G
FERRIS, JFFFFRY
FIELDS, STEVEN H
FILKINS, GIA L
FINEN, ALLEN L
FINNESSEY, GAIL M
FIORITO, ARMOND
FIREHAMMER, FREDRIC S
FISCHER, RODNEY J
FISHER, JACK A
FISHER, LISA M
FISHER, MARK S
FISHER, ROY L JR
FISHER, TIMOTHY R
FISHER, WILLIAM
FITZGERALD, DAVID B
FITZGERALD, RICHARD
FLAHERTY, JOHN F
FLAHERTY, THOMAS M
FLANDERS, FRED A
FLETCHER, KEITH R
FLETCHER, RICHARD E A II
FLICK, TIMOTHY G
FLINTON, TROY L
FLORENCE, PAUL M
FLORES, REX D
FLORES, ROBERT L
FLOWERS, BRIAN S
FLYNN, HUGH B
FODREA, ORMOND R
FOGERTY, JAMES P
FONDULIS, JOHN A
FORD, CHARLES L
FORD, JOSEPH F JR
FORD, KEVIN A
FORD, STANLEY JR
FORNANDER, CHRISTINA M
FORNISS, RONALD E
FORONDA, OSCAR C
FORSTER, HAROLD F
FORTE, MABLE
FORTIER, JON F
FORTIN, JOSEPH L P
FORTIN, PHILIP I
FORTSCHNEIDER, MICHAEL R
FOSTER, GARY L
FOSTER, ROBERT M
FOSTER, VICTOR D
FOUNTAIN, JOHN
FOURNIER, MICHAEL J
FOWLER, DONALD E JR
FOWLER, JOSHUA D
FOX, BRIAN O
FOX, JAMES R
FOX, ROBERT T
FRANCE, JEROME W
FRANCE, NEAL B
FRANCIS, MICHAEL B
ERANCIS, MICHAEL S
FRANCOIS, BERTELL
FRAZIER, ANTONIO
FRAZIER, EDWARD L
FRAZIER, FLOYD L
FRAZIER, JEREMIAH N
FRAZIER, STEVEN W
FREDERICK, ROBERT C
FREEDMAN, MORTON
FREEMAN, BRUCE M
FREIHAGE, RANDALL P
FRENCH, DONALD B
FRENCH, SCOTT W
FREY, DOUGLAS W
FRIES, MICHAEL P
FRIESON, VENORA Y
FROCK, JEANNIE E
FROCK, JOHN C
FROEHLING, R F
FUCHS, DAVID W
FUGIEL, DAVID J
FULLERTON, JAMES A
FULLGUM, DANIEL P
FUNKHOUSER, CHARLES L JR
FURTADO, PAUL J M
FUTCH, ARTHUR E JR
GABBERT, GRANT M
GABLE, BOB L
GABOR, KENNETH M
GADOW, WILLIAM H
GADSON, MICHAEL L
GAENG, WILLIAM J
GALATAS, LONNIE S
GALE, RONALD L
GALINDO, JESSE JR
GALLARDO, JOEL C
GALLOWAY, KERRY L
GALVEZ, ANTHONY P
GAMAB, MARIO E
GANNON, ERIN C
GARABILES, JOSE G
GARCEAU, NORMAN J
GARCIA, ALFRED R JR
GARCIA, DANIEL M
GARCIA, JEROME L
GARCIA RICARDO D
GARLAND, HUGH A

GARRETT, SHEILA K
GARRETTE, JACK L
GARRINGER, WILLIAM J
GARRISON, LON B
GARTHWAITE, RUSSELL E
GASKEW, TONY
GATTUSO, PATRICK R
GAVAGAN, VINCENT J III
GEBERT, RICHARD C
GEIL, FRANZ G III
GEMENES, ANTONIO J
GEMENES, BRIDGETTE
GEMPIS, VALENTINO
GENERETTE, DAN A
GEORGE, JAMES B
GEORGE, TROY A II
GIBB, WILLIAM E
GIBBONS, WILLIAM E
GIBBS, DANIEL R
GIBSON, ALVIN
GIGLIOTTI, ROBERT J
GILBERT, DAVID L
GILBERT, ROBERT C
GILL, JAMES A
GILLAM, JAMES P
GILLEY, MICHAEL T
GILLIAMLEE, JOCELIN O
GILSTRAP, SCOTT E
GILVIN, MARSHALL J
GIROD, SAMMIE R
GIRTEN, JOHN F
GIVENS, KEITH M
GLADDEN, THOMAS
GLEASON, JOE M
GLEASON, KENNETH E
GLENNON, PATRICK M
GLIDDEN, ROBERT M
GLISSON, RICHARD D
GLOVER, DONALD R
GLUTTING, ROBERT J
GOEBEL, DAVID S
GOELITZ, DAVID R
GOLDFARB, BRIAN G
GOMES, JOSEPH N
GONZALES, DAVID J
GONZALES, KENNEDY
GONZALES, MARK A
GONZALES, ROBERT
GONZALES, RODOLFO
GONZALEZ, DUANE B
GONZALEZ, JOSE M
GONZALEZ, RENATO A
GONZALEZ, RITO T
GONZALO, EDDIE D
GOODMAN, JAMES W
GORMAN, BRUCE T
GOTTE, PATRICK J
GRABIANOWSKI, DENNIS E
GRAF, DONNA M
GRAHAM, BRIAN
GRAHAM, DENNIS E

GRAHAM, EDWARD J
GRAHAM, TERRY S
GRANDSTAFF, SCOTT A
GRANT, RAYMOND A
GRANTHAM, DAVID D
GRAUE, WALTER H
GRAVES, MITCHELL J
GRAY, GARY G
GRAY, LAWRENCE T
GREEN, CHARLES B
GREEN, JONATHAN F
GREEN, RAY A
GREENLEE, M
GREENWOOD, WARREN T JR
GREGORIO, MYRNA
GREGORY, LEE A
GREGORY, WILLIAM C
GRESSLEY, RAYMOND E
GRIEGO, TONY A
GRIFFIN, BRYON D
GRIFFITH, RICHARD K
GRIGGS, RAYVON L
GRILLEY, VINCENT
GRIM, REGINA L
GRIME, JEFFREY R
GRIMMER, WILLIAM M
CROSSERHODE, MICHAEL A
GUAJARDO, ERNESTO B
GUBITOSI, ADAM J
GUERRA, RUBEN R JR
GUEST, JOSEPH W
GUIDRY, MELVIN J
GUILL, RICHAFLO R JR
GUILLERMO, ARNIE O
GUILLET, MICHAEL R
GUILLORY, CLARENCE P JR
GUINN, CHARLES J
GUIREY, BUTCH O
GUIRSCH, TODD M
GUNHUS, HELMER A
GUNNER, LEONARD A
GURR, ROBERT N
GUSTIN, GARY L
GUTIERREZ, CAROL E
GUTZENBACH, WILLI
GUZMANPEREZ, JOSE A
HADFIELD, RAYMOND R
HAGER, TIMOTHY J
HAGOOD, PAUL D
HAHN, BRYAN L
HAHN, STEVEN T
HAINES, JOHN A
HALL, ALMA
HALL, CHARLES R
HALL, CHRISTOPHER T
HALL, TEDDY D
HALTLI, THOMAS A
HAMADOCK, BRYAN E
HAMILTON, BRUCE E
HAMILTON, DAVID S
HAMILTON, STEVIE B

HAMILTON, TOM A
HAMMOND, MARK T
HAMMONDS, MITCHELL D
HANCOCK, ANDREW
HAMCOCK, LADDIE K
HANEY, NICHOLAS E
HANKINS, DAVID A
HANKINS, GERALD L
HANKINS, RAYMOND
HANNEMAN, MARK S
HANSEN, GARY T
HANSING, KENNETH R
HANSON, DOUGLAS H
HAPNEY, DAVID S
HARDING, PATRICK T
HARDWICK, ROBERT W
HARGRAVE, ROBERT L JR
HARLIN, KAREN M
HARMENING, RAY D
HARMON, DANIEL W
HARMON, JERRY R
HARMON, MICHAEL L
HARMON, ROBERT D
HARRIS, DENNIS A
HARRIS, ERROL
HARRIS, GAIL P
HARRIS, GREGORY E
HARRIS, LEE N
HARRIS, MICHAEL
HARRIS, PADRAIC A
HARRIS, RI CHARD S
HARRIS, RONALD W
HARRIS, WILLIE A
HARRISON, BRIAN
HARRISON, MALCOLM T
HART, ALAN G
HART, KENTON L
HARTMAN, DONALD C
HARTMAN, LAWRENCE K
HARTNESS, DAVIS D
HARTWELL, ROBERT W
HARTZ, JOHN R
HARVEY, ALBERT B
HARVEY, PATRICK M
HARVEY, ROBERT
HARWOOD, JESSICA E
HATFIELD, PATRICK H
HAUF, ERIC E
HAUGEN, JAMES J
HAUPT, BRADLEY M
HAVARD, DAVID J II
HAVRILLA, JOHN G
HAWKS, MARK W
HAWMAN, CHARLES P
HAWS, JEFFERY L
HAYES, JEFFREY C
HAYNIE, FELECIA A
HAYS, GLYNN T
HAZELTON, WILLIAM
HEARVY, TIMOTHY A
HEDDEN, MICHAEL W

HEDTKAMP, JAMES A
HEFFERNAN, THOMAS J
HEIDLAGE, JOEL A
HEINEMANN, DAVID G
HELMIC, ROBERT F
HENDERSON, JOHN C II
HENDRICKS, HOWARD L
HENDRICKS, RICHARD J
HENDRICKSON, JAMES W
HENRY, JEFFERY A
HENRY, KENNETH L
HENRY, WAYNE G
HENSLEY, BRIAN W
HENSON, BERNARD W JR
HERLIHY ARSENIO S
HERMAN, VICTOR P
HERN, DENNIS O
HERNANDEZ, C A
HERNANDEZ, FERNANDO R
HERNANDEZ, FRANSISCO
HERRERA, DEAN A
HERRICK, DON A
HERRIN, ELBERT A
HERRMANN, PHILLIP J
HESS, LEON E
HESS, RICKY L
HESSLING, KORY A
HETLING, DONALD E
HICKEY, JAMES L
HICKINGBOTTOM, W P
HICKS, ALBERT D
HICKS, MAURICE V
HICKS, REGGIE T
HIGGINBOTHAM, D A
HIGGINS, DERMOTH K
HILER, MICHAEL D
HILL, DIONNA H
HILL, GARY W
HILL, RANDALL D
HILL, THOMAS O
HILL, VONDA K
HILLHOUSE, DAVID J
HILTON, HARRY E JR
HINDLE, BRYAN P
HISH, LLOYD R
HIXON, DARRELL
HOAGLUND, GARY A
HOBBY, JAY A
HODGE, JEFFREY J
HOFFMAN, JOSEPH R
HOFFMANN, MICHELLE J
HOFSOMMER, RICKI C
HOGUE, STEVEN N
HOLDEN, BENTON F III
HOLDEN, STEVEN V
HOLGUIN, ABEL C
HOLLAND, DAVID A
HOLLINGER, JAMES E
HOLLORAN, JOSEPH H
HOLLOWINSKI, DON
HOLMES, JOHN L II
HOLSBERGER, JAMES T
HOLSEY, WILLIE B JR
HOLT, RICHARD C JR
HOLVE, BRADFORD D
HOLZINGER, JAY J
HONHART, DALE T
HONKEN, BRENT A
HOOKS, VIKKI M
HOOPER, LEE E
HOOTEN, PERRY A
HOOVER, ROBERT L
HOPKINS, DWAYNE D
HOPPER, TREVOR I
HORN, STEVEN M
HORN, TONY BRIAN
HORNER, JAMES A
HORNER, RONNELL A
HORSEFIELD, C R
HORSLEY, DOYLE B
HORTIZUELA, GLEN A
HOSTETTER, ANTHONY G
HOUCHIN, RANDY W
HOWARD, FRANK S
HOWARD, FREDERICK V
HOWARD, REGINALD E
HOWARD, WALTER
HOWELL, CHARLES E
HOWELL, GARY R
HOWELL, PETER G
HOWER, PETER C
HOWIE, SPENCER J
HROBAR, RANDOLPH A
HUBBARD, TYRONE A
HUCKABY, RICHARD T
HUCKEBY, ANGELA N
HUDAK, ROBERT N
HUERECA, PHILLIP
HUGHES, DAVID M
HUGHES, JAMES A
HUGHES, JUSTIN D
HUGHES, RANDY L
HUHN, RICHARD S
HULL, PAUL R JR
HUMPHFERS, GREGORY D
HUNT, ERA K
HUNT, ROBERT L
HUNT STEVEN K
HUNT, WILLIAM A
HUNTER, CHARLES H
HUNTER, MICHAEL J
HUNTER, TYRONE T
HUNTSMAN, JABE B
HUNTZ, RICHARD J JR
HURST, JOSEPH L
HURT, THOMAS E
HUTCHINSON, SHAWN A
HYATT, JAMES
ILAOA, ALBERT
IMOTAN, EDGARDO A
IMPERIAL, ULYSSES C
INOUYE, BRIAN
IPOCK, BRADY C
IVEY, CHARLES E
JACKS, ROBERT JR
JACKSON, ELIZABETH L
JACKSON ERIC M
JACKSON, JOHN B
JACKSON, KARIN G
JACKSON, RANDY F
JACOBS, CHARLES D
JACOBSEN, COREY T
JAGIELSKI, MICHAEL P
JAKUBS, CHARLES E JR
JAMES, RONNIE D
JAMES, WILLIAM G
JAMOTILLO, VIRGILIO
JANICKY, JOHN P
JAURIGUI, ESTEBAN H
JAVIER, CRISTITUTO Z
JAVINAR, RICARDO S
JAYCOX, THOMAS P
JENCO, JAMES J
JENKINS, AUSTIN A
JENKINS, JEFFREY S
JENKINS, JOSEPH D
JENKINS, LAURENCE A
JENKINS, MARK A
JENSEN, GARY M
JERKINS, GREG D
JETER, BARRY JR
JIMENEZ, DANIEL
JIMENEZ, MOISES
JINKS, BOBBY J
JOERRES, MARC
JOHNSON, ANGELA F
JOHNSON, BILLY G
JOHNSON, DARYL A
JOHNSON, DOUGLAS D
JOHNSON, ELMER E
JOHNSON, JEFFREY A
JOHNSON, KEITH D
JOHNSON, KEITH E
JOHNSON, KRAIG S
JOHNSON, KYLE S
JOHNSON, RAYNOND H
JOHNSON, ROBERT A
JOHNSON, THOMAS A
JOLIVETTE, GERARD
JOLIVETTE, STEVEN
JONES, ANTHONY B
JONES, BERNARD
JONES, BOYD F W
JONES, BRUCE L
JONES, BRYANT L
JONES, DANNIE N
JONES, DAVID L
JONES, JAMES W
JONES, KEVIN S
JONES, PIERCE L
JONES, ROSE H
JONES, RUSSELL
JONES, STANLEY C

JONES, STEFANY D
JONES, WAYNE E
JONES, WILLIAM E
JORDAN, DARRYL Z
JORDAN, JOSEPH J
JORDAN, MICHAEL
JORDAN, SAMUEL J
JORDAN, WILLIAM A
JORGENSON, JAMES R
JOYNER, JOSHUA R
JOYNER, LANCE L
JOZWIAK, THOMAS
JUAN, MICHAEL S
JUDD, ERIC C
JUDD, JASON W
JURNA, CHARLES V
KANDOLL, MICHAEL
KANE, TIMOTHY N
KANNADY, DOUGLAS N
KAREL, STEPHEN B
KARLBURG, CATHY
KAUFFMAN, TERRY L
KAY, JOEL D
KEATON, ALLAN
KEATON, DAVID
KEIFER, EDWARD J
KELIIKOA, JOSEPH K JR
KELLER, DENNY
KELLY, DARLA J
KELLY, RANDAL G
KEM, MICHAEL
KENDRICK, ROBERT L
KENNEDY, MICHAEL D II
KENNEDY, PAUL D
KENNEDY, WILLIAM A
KERCHNER, KURT N
KERMER, WOLFGANG H
KERN, MICHAEL J
KERR, DWAYNE E
KETNER, KEVIN J
KETTLER, ROBERT J
KEY, ARTHUR JR
KIENZLEN, MICHAEL L
KIESLER, JOHN F
KIEWSKI, MICHAEL D
KIMBLE, CHARLES P
KIMBLE, JAMES H JR
KIMBROUGH, A
KINCAID, DARREN L
KING, BRIAN M
KING, MARK
KING, PATRICK W
KING, ROBERT L JR
KING, RONALD L
KINGSBURY, KENNETH H
KINMAN, SAMUEL D
KINNEY, M A
KINNEY, ZACHARY Z
KINSEY, DONALD F
KIRKPATRICK, DON R
KIRKSEY, ROBERT S
KIRWAN, JON G
KISER, CASE R
KISNER, JANIE L
KISNER, ROBERT E
KITTREDGE, JAMES F
KLEINSMITH, GERARD A
KLINEDINST, RAYMOND F
KLUESNER, CHRISTOPHER M
KMET, JUDY
KNOCHE, LARRY R
KNOLL, JAMES R JR
KNOX, DANIEL G
KNOX, J E
KNOX, WILLIAM J
KOCHEL, LORI A
KOCHER, RICK P
KOENIG, PAUL K
KOPPLER, DANA L
KORBAKES, PHILLIP N II
KREIFELS, SUSAN C
KRISHER, NEAL C
KRIZAN, KELLY J
KROGWOLD, ROGER A
KRUEGER, JOSEPH J
KRUPA, JAMES W
KUBICK, CHRISTOPHER J
KUNTZ, TRACEY C
KURZ, JULIAN T
KUYPERS, CAMERON A
LACOUR, LARRY G
LAFERRELL, SHERIDAN R II
LAFFERRE, MATTHEW A
LAFOUNTAIN, JOHN A
LAFRANCE JEFFREY J
LAGAREJOS, ALEJANDRO L
LAHY, DAVID L
LAICHAS, GEORGE D
LAIRD, BRENT E
LAKE, MILLER G
LAKE, RICHARD J
LAMARCA, RAUL G
LAMB, MARK C
LAMB, MICHAEL J
LAMMERT, EUGENE T
LAMOY, LEON E
LANCIANO, ERIC T
LANE, RICHARD W
LANEY, PAUL A
LANIER, MICHAEL R
LANNON, HOWARD R
LAPIERRE, MARCIA
LAPOINTE, ROBERT L
LAPOINTE, TRACY L
LARGE, RUSSELL D
LARGETEAU, SUSAN R
LARRY, JOHNNIE
LARSON, GEORGE C
LARSON, KARL A
LARSSON, BRIAN D
LASSITER, KEVIN B
LATHAM MARY K
LATHAM, WILLIAM I JR
LATIN, ERNEST
LAURENT, JAMES A
LAVELLE, STEPHEN M
LAVIELLE, MICHAEL S
LAWSON, DAVID M
LAWSON, MARGARET Z
LAZORIK, JOHN L JR
LEAL, JOSE
LEANO, MARCIAL D
LEANO, RICARDO G
LEAVELL, JAMES V
LEBIEDZ JOSEPH G JR
LEBLANC, MARK L JR
LEBUTT, BRIAN W
LEDDING, JON E
LEDOM, JACK L JR
LEE, BRIAN E
LEE, JERRY W
LEE, RUSSELL E JR
LEE, WILLIAM M JR
LEE, WILLIE L
LEE, WOODROW
LEFEVER, MARIE Y
LEFTWICH, BILLY L
LEGRONE, JOSEPH A
LEHMAN, SEAN A
LEITCH, SCOTT A
LEMIEUX, JOSEPH R
LENCZUK, STEVE A
LERCH, RONALD E
LESANE, TERRENCE
LESSNER, LEONARD J
LEVAY, STACY E
LEWIS, BRIAN S
LEWIS, CLARENCE E
LEWIS, DAWNEEN A
LEWIS, DENVER L
LEWIS, DOROTHY
LEWIS, JOEL L
LEWIS, KENNETH S
LEWIS, ROBERT J
LEWIS, ROGER V
LEWIS, ROSE M
LEWIS, TOMMY L
LEY, MIGUEL A
LEYSON, THEODORE R
LIBARIOS, ARNEL C
LIBERT, RODNEY L
LIENHART, PATRICK F
LIGERALDE, GREG D
LIGHTFOOT, JOHN W
LIGON, FLORDELIZA H
LIKINS, GLENNON C
LILLIS, MATTHEW C
LILLY, JOHN E
LINDSAY, VYNA I
LINDSEY, BRIAN D
LINDSEY, JAMES C
LINER, DARRYL D
LITTLE, G

LITTLE, HAROLD D
LITTLE, SUNYATA K
LIVICA, ARISTON P
LLOYD, JAMES W
LOBBESTAEL, ROBERT A
LOCKETT, MELVIN S
LOEWENHAGEN, RICHARD J
LOHMAN, KATHLEEN M
LONCAR, CRAIG P
LONG, DONALD C JR
LONG, GERALD L
LONG, PATRICIA A
LONG, ROBERT
LONG, STEVE
LONG, SUE E
LONGSTREET, RENEE L
LOOMIS, ROBERT M
LOPEZ, DAVID A
LOPEZ, LISA A
LOQUIST, GREGORY S
LORCH, ROBERT E
LORIMOR, RODNEY J
LOTER, KEITH D
LOTT, REGINALD D
LOUDENSLAGER, WILLIAM E
LOUM, GLENN M
LOUSER, BRIAN L
LOVATO, MARC L
LOVE, JAMES E JR
LOVELL, RONALD
LOVETT, BENJAMIN
LOWE, CHRISTOPHER A
LOWE, ROBERT D
LOY, STEVEN W
LOZANO, ALFRED E
LOZINSKI, WILLIAM H
LUCAS, ALEXANDER J
LUCAS, JOHN P
LUCAS, NESTOR P
LUCERO, PETER B
LUCY, JAMES W
LUDINGTON, BRUCE S
LUECHTEFELD, MARK E
LUKE, MARC A
LUNDGREN, ROBERT F
LUPTON, REBECCA R
LUSHER, BRYAN
LUTHER, DALE A
LYDSTON, JEFFREY P
LYLES, JENNIFER R
LYNCH, ANTHONY W
LYNCH, WILLIAM J E
LYNDE, REX K
LYONS, EDWARD
MABE, JOE B
MABUTI, GIL B
MACELUCH, THOMAS D
MACHETTE, ROBERT G
MACK, LARRY J JR
MACKEY, JOSEPH P
MADDAGAN, ROLAND Q
MADERE, DAVID W
MAERZ, JOHN M
MAGANA, SALVADOR
MAGGARD, MARK D
MAGNAN, JOHN J
MAGNERS, DONALD T
MAJORS, ROBERT W
MAKER, ROY
MALLOY, JAMES D
MALM, ROBERT L
MANABET, E
MANALO, JUN L
MANDL DAVID
MANECKE, WALTER J
MANGLONA, FRED C
MANINO, DAVID A
MANN, JAMES T
MANNING, DANNY W
MANNING, GEORGE A
MARCH, JEFFREY E
MARCOS, ISIDRO E
MARIANO, ENRIQUE
MARINELLI, BRIAN K
MARKEY, KEITH E A
MARKHAM, PAUL E
MARQUEZ, GLENN
MARQUEZ, HERMIE C
MARS, GEORGE R JR
MARSH, HAROLD J
MARSHALL, KEVIN B
MARSHALL, RENE H
MARTIN, ARTIST H
MARTIN, DARREN W
MARTIN, DAVID M
MARTIN, JAMES E JR
MARTIN, JEFFERY A
MARTIN, ROBERTA L
MARTINEAU, DONAVON J
MARTINELLI, MARY M
MARTINEZ, ALBERTO
MARTINEZ, LOUIS H
MARTINEZ, MARK R
MARTYNOVYCH, PETER N
MARYNEN, SIMON
MASE, STEPHEN A
MASSEY, DEON M
MATHENY, SCOTT A
MATHEY, GUY E
MATHIS, JOHN H
MATHIS, MELINDA F
MATTHEWS, LINDA R
MATTINGLY, BRIAN F
MAULDIN, TROY A
MAURER, DOUGLAS A
MAXEY, CHRISTOPHER
MAXWELL, JAMES R
MAY, CLEODIS
MAY, FLOYD
MAY, RICHARD L II
MAYER, DWIGHT J
MAYES, ROBERT L II
MAYNARD, MICHAEL D
MAYNARD, YETTA L
MAYORAL, LUIS A
MAZZONI, ANTHONY J
MAZZONI, THOMAS J
MCALINEY, JOHN P
MCALISTER, KIMBERLY K
MCALLISTER, THOMAS K
MCBRIDE, LEWIS E III
MCBRIDE, MICHAEL A
MCCALL, ALBERT T JR
MCCALLUM, WILTON E
MCCARTER, RONALD A
MCCARTHER, DALE R
MCCARTHY, DENNIS S
MCCARTHY, EDWARD J
MCCLADDIE, CURTIS
MCCLEES, DOUGLAS M
MCCLELLAND, SHAWN E
MCCLINCHY, TIMOTHY J
MCLOUD, DAVID L JR
MCCLUNG, JAMES L
MCCLUNG, ROCKE W
MCCLURE, JEFFERY
MCCLURE, RICHARD W
MCCORMICK, KEVIN M
MCCORQUODALE, CRAIG
MCCORT, BRADLEY R
MCCOY, CARL H JR
MCCOY, LEONARD C
MCCRORY, JOHN P
MCCULLOUGH, EDWARD F
MCCUTCHEON, JOHN E
MCDANIEL, HOSHEA M
MCDERMOTT, MARK J
MCDONALD, HAROLD R
MCDONALD, WILBERT JR
MCGACHEY, JAMES E
MCGANN, LISA L
MCGEE, MARK V
MCGILLIVRAY, WARREN
MCGRIFF, ROBERTA L
MCGUIRE, DON J
MCGUIRE, KEITH A
MCGURK, EDWIN W
MCHENRY, EDWARD H
MCINTYRE, MICHAEL L
MCKAY, DAVID A
MCKEAN, KURT D
MCKELLIPS, KYLE B
MCKENZIE, ROBERT
MCKEY, JOHN M
MCKINNEY KRIS F
MCLAIN, SAMUEL A
MCLAUGHLIN, MICHAEL W
MCMANIGAL, WILLIAM K
MCMANN, CHARLES R
MCMILLAN, BRIAN N
MCMILLER JEFFERY L
MCMURTRAY, THEODORA
MCNAIR, DALE D

MCNEELY, ROBERT L
MCNEIL, WAYNE C
MCWHIRTER, MARK E
MEADOWS, KERRY E
MEANLEY, PAUL
MEANS, MARY J
MEANS, WARREN D
MEDINA, ALVIN G C
MEDINA, LEONARD III
MEEKER, ROBERT W
MEIER, DEBORAH K
MELTON, PAUL F
MENDEZ, GAVINO J
MENDOZA, ANTHONY C
MENDOZA, MAYNARD M
MENDOZA, RAYMUNDO C
MENOR, REYNALDO L
MERCENDETTI, C
MERTENS, DAVID A
METSGER, MARK A
METTING, PAUL A
MEYERS, ERIK J
MICHAEL, DANIEL L
MIDEL, DAVID L
MIKKELSEN, MICHAEL L
MIKKELSEN HARLAN E
MILANOVITS, GUY L
MILBURN, MICHAEL D
MILES, DAVID A
MILLARD, LAWRENCE S
MILLER, BENJAMIN M
MILLER, CHARLES E
MILLER, DAVID L
MILLER, DEAN A
MILLER, DEAN J
MILLER, JAMES W
MILLER, JOHNNIE L
MILLER, JOSEPH M
MILLER, KENNETH L
MILLER, LEON D
MILLER, MICHAEL G
MILLER, STEPHEN R
MILLER, STEVE
MILLER, WILLIAM L
MILLER, WILLIAM S
MILLS, JOHN
MILLS, YASUTOMO
MILNER, RICHARD W
MILTON, RANDY L
MINA, ROLANDO L
MINCY, CHRISTOPHER W
MISA, SERGIO M M
MITCHELL, BILLY G
MITCHELL, DANIEL T JR
MITCHELL, DONALD J
MITCHELL, PATRICK O
MITCHELL, STEPHEN C
MOEN, DANA S
MOFFETT, JOSEPH C
MOFFETT, MARCUS S
MONROE, GARRY A
MONTANO, MANUEL
MOON, RICHARD A
MOON, RICKY D
MOORE, BOBBY R II
MOORE, FRANK L
MOORE, JOHN D
MOORE, MICHAEL A
MOORE, TIMOTHY J
MORALES, VICTOR A R
MORELOS, ALBERT O
MORGAN, ALAN L
MORGAN, DENNIS D
MORGAN, JAMES P JR
MORGAN, KIRK
MORGAN, PATRICK J
MORGAN, ROBERT J JR
MORGAN, THOMAS P
MORREY, SHAWN C
MORRIS, BOBBY D
MORRIS, BRYAN A
MORRISON, JAMES M T
MORRISON, JOSEPH S
MORRISEY, ROBERT P
MORSE, JAMES G
MORTON, WILLIAM
MOSER, KENNETH J
MOSES, JOHN J
MOSHER, ROBERT E
MOSIER, GARY L
MOSS, DAVID A
MOSS, ROBERT B
MOTAKA, CRAIG G
MOTT, DONALD V
MOULTON, GREGORY S
MOYER, BARTON J
MOYER, WILLIAM A
MOYNIHAN, JAMES D
MULLINAX, HARRY K
MUNN, TERRILL R
MURDOCH, GEORGE L
MURISON, CHARLES N
MURPHY, GEORGE R
MURPHY, LOUIS M
MURPHY, MICHAEL P
MURRAY, ANDREW E
MURRAY, KEITH A
MYERS, LAWRENCE F
MYERS, WESLEY R
MYRICK, CEDRIC
NADEAU, TIMOTHY J
NAFRADA, CAESAR E
NALLS, PAM
NAPIER, ROBERT B
NASIELSKI, TIMOTHY
NATIVIDAD, DANNY P
NAZARENO, DENNIS T
NEAL, ROBERT E
NEECE, LESLEY A
NEGRON, RAY R
NELSON, CURTIS
NELSON, JAMES A
NELSON, JAMES R JR
NELSON, KELLI L
NELSON, STEVEN J
NELUMS, RICK
NEMEDY, JOSEPH A
NESS, MARLIN A
NESTER, GEORGE W
NEUBECK, PATRICK A
NEVILLE, KIMBERLY K
NEWBERRY, BOBBIE L
NEWELL, DEREK T
NEWMAN, EDWARD M
NEWMAN, PAUL E
NEWSON, HARVEY D
NEYPES, NATHANEAL C
NGO, THANH T
NGUYEN, T V
NICELY, WILLIAM C
NICHOLLS, GERALD W
NICHOLS, CHARLES G
NICHOLS, CHARLES M JR
NICHOLS, CHARLES R
NICHOLS, WAYNE E
NICHOLS, WILLIAM D
NICHOLSON, WILLIAM A
NICKLOW, DAVID A
NICODEMI, MICHAEL S
NIENKAMP, STEVEN J
NISSON, KEVIN G
NIVEN, GEORGE W JR
NOEL, GILBERT
NOONAN, MARK J
NORDSTROM, CARL D
NORDSTROM, JUDITH A
NORMAN, DAVID
NORRIS, EDWARD G JR
NORSWORTHY, STEPHEN B
NORTH, JAMES J
NORTHRUP, THEODORE R
NORTON, BRADLEY P
NORTON, KELBEY
NULL, JAMES R
NUTTER, GLENN P
NYANDER, BRET L
OAKLEY, JEFFREY N
OAKLEY, JIMMY R
OBERMEYER, GARY W
O'BRIANT, JOHN D
OBRIEN, EUGENE C
OBRIEN, JOHN A
OBRIEN, ROBERT F JR
OBRIEN, SEAN P
OCAMPO, DANILO T
OCASIO, ISRAEL T JR
ODA, BRENT M
ODORCICH, THOMAS M III
ODRA, MARTIN A
OIE, DALE R
OILAR, MICHAEL S
OKELLEY, BRADLEY E
OLAES, ROBIN E

OLIVAS, TOMMY
OLIVER, DAVID A
OLSEN, GEOFFREY T
OLSEN, JOHN A
OLSON, THOMAS E
ONEIL, ERIC T
ONEILL, CARL W
ONEILL, CRAIG E
ORDAL, JAMES G
ORESKOVICH, JOSEPH T
ORG, HARRY
OROURKE, JOHN P
OROYAN, PEDRO
OROZCO, STUART W
ORTEGA, DOMINGO JR
ORTIZ, RUBEN
OSBORNE, DERICK M
OSTRANDER, DONALD C
OTOOLE, GORDON G
OWEN, KELLY T
OWENS, CHRISTIAN M
OWENS, LARRY G
OWENS-HUNT, LISA S
PACIELLO, RONALD
PACK, JOSEPH B
PACKARD, JERRY M
PADILLA, ERIC T
PAINE, LARRY D
PALAIS, LAURENCE
PALECEK, DOUGLAS A
PALM, CHRISTOPHER M
PALMER, DALE I JR
PANKOP, JEFFREY P
PAPP, JOHN T
PARDE, CHAD A
PARK, ROBERT
PARKER, GEORGE G
PARKER, GREGORY P
PARKER, HUGH
PARKER, ISRAEL A
PARKER, JOHN C
PARKER, MICHAEL L
PARKER, RAYMOND L
PARKER, ROBERT M
PARKMAN, ALLEN K
PARKS, ALEX J
PARKS, JUDITH
PARRISH, MARK G
PARVIN, NORMAN D
PASCAL, FREEMAN P IV
PASSWATER, THOMAS A
PATTERSON, JEFFREY J
PATTON, ROBERT J II
PATUBO, PERCIVAL M
PAVEGLIO, DAVID A
PAXSON, KELLY R
PAYNE, DALE
PAYNE, EDWARD E
PAYNE, JEFFREY C
PAZ, EDUARDO V
PEACOCK, JAMES R
PEARSON, ANDREW R
PEARSON, ARTHUR J
PEARTREE, KENNETH W
PECK, JAMES R
PECK, JERAULD F JR
PECK, ROBIN A
PEDDY, JOHN L
PEET, NORAH J JR
PEGG, HARRY T JR
PEHOWIC, PATRICK A
PELLICHET, MICHAEL A
PEMBERTON, FORREST T
PENA, JACINTO
PERALTA, ARTHUR D
PEREZ, DAVID G
PEREZ, DAVID L
PEREZ, GILBERT A
PEREZ, RICHARD L
PERKINS, JEFFERY A
PERRINO, DANIEL
PERRINO, DIDI
PERRY, JAY M
PERRY, LEOZIE
PERRY, MICHAEL C
PERRY, RONALD J
PETERS, ANDREW M
PETERS, PHILIP N
PETERS, THOMAS C
PETERSEN, BRANDON G
PETERSON, KIMBLE L
PETERSON, LAWRENCE F JR
PETRI, KEVIN C
PETTIT, TROY A
PHELPS, MICHAEL F
PHILLIPS, ANTHONY M
PHILLIPS, CHARLES E
PHILIPPS, KENT L
PHISAYAVONG, S
PIARROT, ERIC W
PICKENS, EDWIN G
PICKERING, ERANK
PICKETT, DALE A
PIERCE, BILL D
PIERCE, GARY W
PIERCE, MICHAEL D
PILA, ROBERT J
PILLING, STACEY A
PINNIX, WILLIAM F
PITTMAN, BARBARA A
PITTMAN, PAUL E JR
PITTMAN, RUSSELL S
PITTS, DAVID D
PLATTER, WILLIAM E
PLEDGER, TIMOTHY C
PLOCHOCKI, THOMAS J JR
PLOTT, JOSEPH
POBLETE, JOSELITO C
POCHEK, JAMES W JR
POLLARD, KEVIN W
POLSTON, CODY F
PONSART, KENNETH A
POOR, LARRY C
POPE, ROBERT G
POPPE, DONALD R
PORTER, JAMES
PORTER, NATHAN K
PORTER, RICHARD
PORTER, WAYNE H
POWELL, KEITH D
POWELL, LARRY D
POWERS, JODY R
PRATT, JEFFREY A
PRATT, VICTOR G
PRECHTER, GARY C
PRENDERGAST, PHILIP A
PRESA, NESTOR A
PRESTON, KELVEN
PRICE, HEWLETT D JR
PRICE, MITCHELL A
PRICE, ULYSSES
PRIES, THOMAS
PRITCHETT, LEE
PROCTOR, DOUGLAS W
PRUDENCIO, JUN C
PRYOR, JOHN E
PUDDU, JOSEPH R
PUGH, MICHAEL
PURVIS, JAMES T JR
PURYEAR, RON
PUTNAM, PAUL J
QUARLES, MICHAEL O
QUIGLEY, DAVID P
QUINGUA, NELSON A
QUINN, STEPHEN W
QUINONES, ISRAEL JR
QUINT, JAMES P
QUINTO, ROMULO C
QUIZON, WILSON S
RABEY, DEREK W
RADEMACHER, LORI A
RAFI, RUEL M
RAGIN, JOHN
RAGLIN, ROBERT F JR
RAINGE, RONNIE
RAINS, KIRK
RAINS, KERRY F
RAMOS, CARLOS G JR
RAMOS, EDWARD C
RAMOS, JAIME L
RAMOS, JESUS
RAMOS, ROLAND V
RAMSEY, ALVIN L
RAMSEY, ROBERT L
RAMSEY, WILLIAM M
RANCK, CHRISTOPHER A
RAND, RONALD T
RANSFORD, PAUL J III
RANSOM, DAVID M
RAUTH, DONALD R
RAWLS, ANTONIA
RAY, KILPATRICK A
RAY, WILLIAM D

RAYBON, MARTY C
REAGAN, RONNIE
REAGIN, MARK A
REASE, JAMES A
REAVES, STEPHEN M
RECINTO, FRANCISCO P
RECK, STEPHEN
RECORE, BRUCE M
REDFORD, IVAN D JR
REDIC, ALBERT L
REDINGTON, ROBERT R
REED, MARK S
REED, MILLAND
REEVES, JAMES H II
REEVES, NORMAN E
REGISTER, ROBERT A
REICH, EDWARD A
REICHERT, JEFFREY E
REID, ALFRED A JR
REIDY, CHARLES, A
RENDLE, JOSEPH W
RENDON, REUBEN A JR
RENKEL, AARON G
RENNER, ROBERT P JR
REYES, RAFAEL A S
REYES, VICTORINO O JR
REYNOLDS, ROB ERT F
REYNOLDS, THOMAS A JR
RHOADES, WAYNE L
RHODES, VERNON L
RICHARDS, EUGENE D
RICHARDS, PAUL E
RICHARDSON, ARNOLD B
RICHARDSON, ERIC L
RICHARDSON, JAMES E
RICHARDSON, LAWRENCE E
RICHARDSON, VIRGIL
RIDDLE, MARK D
RIDENHOUR, WILLIAM C
RIDER, JONATHON C
RIDER, JOSEPH C
RIDGEWAY, LARRY L
RIGGLE, STEVEN A
RIGGS, TIMOTHY R
RIGUAL, RITA D
RILEY, HUGH H
RINARD, JEFFERY L JR
RISTOW, KEITH W
RITCHEY, STEVEN R
RITTER, GREGORY H
RITTGERS, MARY J A
RIVARD, ANTHONY G JR
RIZZO, ANTHONY G JR
ROACH, CHARLES W
ROBERTS, DANNY C
ROBERTS, DAVID W
ROBERTS, DEAN E
ROBERTS, JACK
ROBERTS, KERRY K
ROBERTSON, J T
ROBERTSON, JOHN O

ROBINSON, BENJAMIN L
ROBINSON, BOBBY L
ROBINSON, CARL
ROBINSON, FRANK
ROBINSON, HEIDI E
ROBINSON, LESLIE J JR
ROBINSON, RAYMOND E
ROBINSON, ROBERT J
ROBINSON, THOMAS E JR
ROBSON, WILLIAM T JR
ROCCO, CARL D
ROCKHOLD, MICHAEL T
RODRIGUEZ, ALFRED
RODRIGUEZ, ALFRED D
RODRIGUEZ, CELESTINO JR
RODRIGUEZ, RICHARD D
RODRIQUEZ, JOSEPH A
ROE, CHRISTOPHER P
ROGERS, GERALD R JR
ROGERS, LARRY H
ROHN, JOHN M
ROLLYSON, EDWARD L
ROMBAOA, MACARIO B
ROMERO, JORGE H
ROMERO, MARIO M
ROOFNER, WILLIAM C
ROONEY, TIMOTHY J
ROPER, TOMMY W JR
ROSA, CHARLES
ROSA, ESPERANZA DC
ROSA, ISRAEL
ROSA, MARLENE S
ROSALES, JOSE D JR
ROSAMOND, E A
ROSE, HENRY L
ROSS, EDGAR L
ROSS, JOHN R II
ROSS, SEAN M
ROSSI, MARK D
ROTH, SHERI J
ROUSEY, GREGORY W
ROUTH, CHARLES L
ROWE, BRADLY D
ROZANSKI, JOHN G JR
ROZANSKI, KEVIN F
RUBIO, G P T
RUDDER, LESTER F
RUIZ, PHILIP J
RUMSEY, THERON F
RUSHING, JOHNNY C II
RUSSELL, EMORY C
RUST, WILLIAM J
SAACK, MORIA C
SABINO, VICTOR L
SACCUMAN, MARIO B
SADORRA CORNELIO T
SAGE, ALLAN J
SALAZAR, EDWARD R
SALAZAR, SANDRA L
SALGADO, MELANIO R
SALSBERRY, DARCEY A

SALZMAN, KENNETH L
SAMMS, LYNN K
SANCHEZ, HARRY J
SANDERS, GEORGE G
SANDERS, JANET L
SANDERS, JUDITH A
SANDERS, RANDAL J
SANDLER, MICHAEL
SANDLER, NANCY A
SANDY, STEVEN E
SANNEY, GARY K
SANTA MARIA, FREDERICK
SANTACROCE, DAVID G
SANTACRUZ, REY C
SANTHOFF, SHERRY R
SANTOS, JESUS
SANTOS, RONNIE V
SARACENO, MICHAEL S
SARAO, JOHN
SARGENT, RANDY D
SATTERWHITE, RONALD L
SAUCILLO, RAMIRO J
SAUNDERS, DUKE C
SAUNDERS, FREDERICK E JR
SAVAGE, DONALD S
SAVANT, DAVID L
SCHAAF, GORDAN B
SCHAEFER, CANDACE M
SCHAEFER, DARRELL L JR
SCHAEFER, DAVID P
SCHAEFER, SUZANNE T
SCHAUB, FREDERICK L
SCHAUM, HAROLD D
SCHEIERN, TERRY L
SCHENKEL, JAMES F
SCHETROMA, KENNETH L JR
SCHIMITT, MARK
SCHLINK, DANNY R
SCHLINK, MICHAEL L
SCHLINK, ROSALIND
SCHMITT, RICHARD W
SCHNEE, THOMAS M
SCHOLTISEK, ERIC V
SCHOMAKER, LAURENCE W
SCHOZER, DAVID A
SCHRIEWER, RICHARD S
SCHROEDER, RANDALL K
SCHROER, CORY L
SCHRUMP, DONNA M
SCHULTZ, RAYMOND H
SCHUTZUIS, MARVIN J
SCHWAB, THOMAS JR
SCHWEIZER, ALVIN C II
SCHWOERER, RONALD E
SCORDIA, FRANCESCO
SCOTT, CHRISTINE J
SCOTT, DEAN M
SCOTT, GLENN M
SCOTT, JONATHON A
SCOTT, ROGER J
SCOTT, TIMOTHY D

SCOTT, TRENT L
SCOTT, WENDY A
SCRIBNER, JAMES B
SCRIVER, DAVID C
SCROGGINS, BERTON S
SCUITO, RICHARD E
SEAGRAVES, TRACY L
SEAMAN, RICHARD L
SEE, PETER T
SEIF, KENNETH
SEITTER, ROBERT L
SELBITSCHKA, DANIEL L
SELDERS, WELDON R
SELLS, DANNY
SEMIEN, LORIE B
SERINO, JAMES P
SESSION, TYREE V
SEVERANCE, PETER G III
SEWELL, JONATHAN S
SEXTON, KURT J
SHAFFER, GEORGE E
SHAMPO, JEFFREY D
SHANKLES, BOBBY D
SHANNON, CHARLES G
SHANNON, RICKY
SHANOR, EDWARD F JR
SHARER, JAMES D
SHARON, CHARLES L
SHARP, DAVID W
SHARP, HARRY R III
SHARPS, DORAL D
SHAW, JEANNIE M
SHEFFIELD, MAURICE
SHELNUTT, RICHARD T
SHEPHERD, MICHAEL A
SHERRILL, STEVEN W
SHERROD, DEMEATRICE
SHIELDS, JUDITH A
SHIELDS, RICHARD K
SHILLING, MICHAEL C
SHIREY, JEFFREY A
SHIRLEY, FRED A
SHOCKLEY, THOMAS E
SHOEMAKER, CLIFFORD
SHOLES, JOSEPH D
SHOVER, MATTHEW T
SHOWALTER, FLOYD W
SHULL, LARRY W
SHURTZ, EDWARD A
SIAPNO, ALFRED C
SIBERT, JAMES H
SIFUENTES, DIEGO A
SILER, JAMES
SILVA, CARLOS L
SILVERMAN, JESSE M G
SIMIC, MICHAEL S
SIMMONS, ALPHONSO
SIMMONS, MARK L
SIMMONS, PATRICK
SIMMONS, PAUL M
SIMON, KEITH E
SINGELTON, JOHN C
SIPES, TIMOTHY E
SKAJEM, CHARLENE R
SKEIDE, DOUGLAS R
SKELTON, JOHN B II
SKINNER, JOHNNY
SKIPPER, MARVIN
SKIPPS, DAVID A SR
SKOWRONSKI, DIANE L
SLACK, CINDY L
SLAGLE, STEVEN A
SLATER, BRIAN L
SLATER, RICHARD
SLIGAR, WILLIAM G
SMITH, ALTHEA A
SMITH, ANNA A
SMITH, BRIAN D
SMITH, BRYAN H
SMITH, BUDDY G
SMITH, CARL L
SMITH, CHARLES
SMITH, CHRISTOPHER J
SMITH, CLIFTON M
SMITH, DANIEL J
SMITH, DAVID G
SMITH, DAVID W
SMITH, DENNIS T
SMITH, DOUGLAS E
SMITH, EARLE G III
SMITH, FRANK M
SMITH, GERALD A
SMITH, GREGORY S
SMITH, JAMES M
SMITH, JOAN C
SMITH, JODY
SMITH, LANCE H
SMITH, LEONARD C
SMITH, LORRIE M
SMITH, LYNN E
SMITH, MARIA A
SMITH, MARK A
SMITH, MARK E
SMITH, MICHAEL C
SMITH, MICHAEL L
SMITH, OWEN D JR
SMITH, ROBERT G
SMITH, ROGER
SMITH, SCOTT A
SMITH, SHARLENE L
SMITH, SHELDON S
SMITH, TIMOTHY D
SMITH, WILLIAM H
SMITH, WILLIE C
SMITH, YVAN K
SNOW, MARK K
SNYDER, ROBERT R
SOBIN, IVAN D
SOBKOVIAK, LAUREN R
SOCEY, MICHAEL T
SOMBKE, WILLIAM L
SOMERS, DANIEL J
SOMMERS, MARC A
SONODA, MICHAEL M
SORIANO, RONALDO T
SOSSAMON, JOSEPH W
SPARKS, WILLIAM F
SPARR, LARRY R
SPEAR, ERIC J
SPEARS, HARRY C JR
SPECK, MICHAEL B
SPEEDIE, JOHN D
SPEER, KEVIN L
SPEIGNER, ROBERT W
SPENCER, EDWARD E
SPENCER ERWIN A
SPILLANE, ALBERT J
SPINAS, DANIEL R
SPRUILL, MILON W
SPRY, JOHN A
SQUIER, SCOTT R
STACEY, GARY B
STACKS, MICHELLE A
STAFF, JOHN K
STAFF, NATALIE T
STAFFORD, DAVID B
STAHELI, RICHARD D
STAMPS, DAVID E
STANFORD, MABLE L
STANKOVICH, ROBERT M
STANSBERRY, TONY E
STAPLES, SHANNON M
STARK, JAY D
STATEN, PATRICK D
STAUFFER, MARK J
STEBBINS, DAVID M
STEELE, ALSTON G
STEELY, ROBERT B
STEINBECK, MICHAEL O
STEINKE, WILLIAM J
STEPHENS, WILLIAM D
STEVENS, CHRISTOPHER L
STEVENS, MITCHELL D
STEVENSON, SEAN T
STEVENSON, TIRRELL D
STEWARD, DOUGLAS E
STEWART, ARVEL
STEWART, BRUCE E
STEWART, CURTIS O
STEWART, STEVEN W
STEWART, THOMAS L
STICHTER, TOMMY E
STICKELS, JAMES E
STINE, NORMA L
STISSER, GARY W
STIVERS, PHILIP T
STIVES, JAMES E
STOGSDILL, JAMES E
STOLL, HAROLD W
STONE, CHARLES D
STORMAN, ROGER A
STOTT, DAVID E
STOUDT, LEE A JR

STRATTON, MARK F
STREPEK, RANDOLPH M
STRICKENBERGER, A L
STRICKLAND, CORTEZ R
STROMBECK, MARK R
STRONG, MATTHEW
STROUD, JAMES R
STROZIER, JOHN C
STUDER, WILLIAM A
SUBER, GARY SR
SUGHRU, SHARON K
SULLIVAN, BONNIE J
SULLIVAN, DONALD N JR
SULLIVAN, JOHN W
SULLIVAN, RANDOLPH L
SUMMERWELL, RODNEY E
SUNDAY, DONALD T
SUSSMAN, JEFFREY W
SUTIONO, ZADOK
SUTTON, DONALD H
SUTTON, SCOTT D
SUTTON, SEAN R
SVATOS, JOSEPH M
SVEDA, DANIEL P
SWAN, STEVEN C
SWANN, DARREN L
SWANSTROM, RICHARD
SWEET, WILLIAM J
SYLVA, KENNETH L
SZPLETT, STEVEN J
TALERICO, JAMES L
TAMBOT, ROQUE B
TAMONDONG, ROMILYN S
TARKINGTON, SAM JR
TATE, STEVEN A
TATUM, DWAYNE
TAUS, THOMAS S
TAWNEY, DANIEL J
TAYLOR, CHARLES C
TAYLOR, CHARLES R
TAYLOR, CHARLOTTE V
TAYLOR, CLINTON R
TAYLOR, DAVID O
TAYLOR, DEAN R
TAYLOR, HAROLD
TAYLOR, HENRY D
TAYLOR, JAMES R
TAZOI, DOUGLAS
TEAS, RONNY R
TEATES, BRYAN C
TEMPLE, ROBERT C
TERRELL, DEBRA L
TERRELL, WILBUR L JR
TERRY, BILLY W
TERRY, GARTH E
TESKE, RANDALL L
TETRAULT, JEANLUC
THARP, SCOTT C
THEALL, JEFFREY S
THOMAS, BERNARD
THOMAS, DAN H
THOMAS, DEXTER L
THOMAS, DONALD L JI
THOMAS, FLOYD R
THOMAS, FREDDY
THOMAS, KENNETH G
THOMISON, WILLIAM
THOMPSON, DAVID K
THOMPSON, GILBERT W JR
THOMPSON, GORDON E
THOMPSON, JAMES E
THOMPSON, KENNETH L
THOMPSON, MARVEL G
THOMPSON, ROBERT F
THOMPSON, WHITNEY L
THOMPSON, WILLIAM B
THORP, RONALD J
THORPE, WILLIAM J
TIBBETTS, ELVIN R
TIEDE, JESSE F
TILLSON, SCOTT R
TIMBOL, LODIVINO C
TINALIGA, BERNADO Z
TINSLEY, DAVID W
TIONG, CONRADO P
TIZON, ROBERTO R
TOBEY, VERNON E
TOLES, MARIA Y
TOLLISON, JOHN K
TOMA, DANIEL R
TOMCZAK, JEFFREY J
TOMKOWSKI, JAMES S
TOMLINSON, JIM K
TONACK, BRIAN
TONACK, ELLEN
TONN, JOHN M
TOPPA, RICHARD A JR
TOPPIN, CARLYLE
TORRENCE, STEVEN G
TORRES, ROBERT J
TORRES, VICTOR W
TOWNSEND, KIMBERLY
TRACE, PHILIP H
TRACY, CRAIG D
TRAHAN, RANDALL J
TRAPP, DENNIS M
RAVAGLIN, GIFFORD L
TREVINO, MARTIN
TRIMBLE, MICHAEL L
TRIPLETT, GARY M
TROTH, MELVIN E
TROUT, BETTY J
TRUITT, HENRY L
TRUMBULL, JOHN J
TRUMP, BOBBY R
TRUMPH, SKIP E
TRYON, IRA K
TUBBS, KENNETH W
TUBIG, DIOSDADO I
TUCKER, DAVID L
TURLINGTON, ROBERT R
TURNER, COREY M
TURNER, MICHAEL L
TURRIST, FRANK D JR
TUSMAN, STEVEN B
UCCI, HAL S
UNDERWOOD, ARTHUR L JR
UNDERWOOD, MICHAEL C
UPDIKE, MICHAEL D
UPP, RONNY L
VALENTINE, ALPHONSON
VAN OMEN, PRISCILLA
VAN ORNE, RONALD W
VANBUSKIRK, ROBERT A
VANCE, ROBERT A
VANDENAKKER, STEPHEN J
VANDERBILT, CHARLES JR
VANDERPOOL, VERNON R
VANGUNDA, CYRUS W
VANNOY, MICHAEL
VANOMEN, VICTOR
VANSELOW, ROBERT J
VANZANDT, GEORGE H
VARDEMAN, MARVIN B
VARHEGYI, JAMES C
VASSER, MEDIATRIX L
VAUGHAN, ROBERT
VAUGHAN, ROBERT E JR
VAUGHN, ROBERT K
VAUGHN, WILLIAM H JR
VAZQUEZ, MARIA C
VELA, ISRAEL JR
VELORE, MARTIN
VENDZULES, MICHAEL C
VENNERI, STEPHEN J
VENTURA, ERNESTO J
VICENS, WILFREDO
VICHA, WAYNE
VIEIRA, CECIL A
VILLAGRACIA, B D JR
VILLANUEVA, ALEX S
VILLANUEVA, MARCEL R
VILLARRUZ, FRANKIE B JR
VILORIA, JOSE V
VISALDEN, EDWIN A
VO, KIM
VOGEL, DOZIER
VOGEL, RICHARD P
VOGT, GARY J
VORWALD, KEVIN R
VRMEER, HOWARD D II
WAKELEY, SHAWN E
WALKER, FRIEND L
WALKER, GARY
WALKER, HAI TRAN VAN
WALKER, LAWRENCE
WALKER, RAYMOND E
WALKER, ROY
WALKER, WELLS W JR
WALL, KEITH E
WALLA, JON T
WALLACE, JOHNNIE L JR
WALLACE, STEVEN J

WALLER, WILBERT
WALTERS, DAVID A
WALTERS, RICHARD E
WARCHOL, GEORGE W JR
WARD, BILLY JR
WARD, DOUGLAS K
WARD, JERRY R
WARE, DON W JR
WARN, CHRISTOPHER L
WARNER, ALFRED R
WARNER, TERRY L
WARNER, TODD M
WARREN, WALLACE C
WASHINGTON, ANDREW
WASHINGTON, CHRISTOPHER A
WASHINGTON, EDWARD D
WASHINGTON, JOHN L
WASHINGTON, WALTER L
WASKO, BRIAN M
WASSON, BARRY G
WATERMAN, SCOTT M
WATERS, MITCHELL D
WATERS, RICHARD B
WATKINS, PAUL L
WATKINS, ROBERT P JR
WATKINS, VANCE V JR
WATSON, CECIL C
WATSON, DONNA E
WEATHERS, MARVIN J
WEAVER, DAVID L
WEAVER, ROGER E
WEBB, JASON D
WEBER, JEFFREY J
WEBERT, MICHAEL J
WEBSTER, MELISSA Y
WEED, JAMES C
WEGMANN, TIMOTHY M
WEGNER, HARRY J
WEHMEYER, MICHAEL E
WEINBURG, PAUL M
WEIR, ERIK D
WELCH, BARBARA J
WELLS, BRIAN A
WELLS, DONALD H
WELLS, MONA
WELLS, ROBERT T
WELSH, WILLIAM J
WENGERTER, ROBERT P
WENNER, STEVEN E
WENZEL, CLIFF G
WERNER, SCOTT J
WEST, ROBERT T
WESTMORELAND, RAY T III
WHIDDEN, RONALD J
WHIRTLEY, BRIAN W
WHITAKER, ALBERT F JR
WHITAKER, CHARLES L
WHITAKER, REAGAN C
WHITE, CHARLES G
WHITE, DANIEL W
WHITE, DAVID M
WHITE, DEKALE
WHITE, THOMAS P
WHITEHEAD, KEVIN
WHITING, JIMMY D
WHITMER, JOHN
WHITT, MICHAEL B
WICHERS, STEVEN M
WICK, HOWARD J JR
WICKIZER, GEORGE W III
WIESE, BRUCE A
WIGGINS, WILLIAM R
WILEY, EDWIN A
WILKERSON, DAN K
WILKINS, SANFORD E
WILKINSON, JAMES E
WILKINSON, KENT G
WILKINSON, RUSSEL
WILLIAMS, BILLY W
WILLIAMS, CARL III
WILLIAMS, CLINT
WILLIAMS, DAN R
WILLIAMS, DANIEL P
WILL IAMS, DAVID M
WILLIAMS, DEREK E
WILLIAMS, ERIC B
WILLIAMS, ERIC S
WILLIAMS, JACCUE P
WILLIAMS, JOHN L
WILLIAMS, PAUL D
WILLIAMS, RANDALL P
WILLIAMS, RODNEY M
WILLIAMS, TYRONE M
WILLIAMS, VERNON R
WILLIAMS, WINFRED
WILLIS, JUNE M
WILLMAN, JAMES D
WILLMAS, MICHAEL J
WILSON, BRIAN C
WILSON, CARL B
WILSON, DAVID I
WILSON, DEAN D
WILSON, DEVIN
WILSON, GEORGE III
WILSON, GLENN J
WILSON, RODNEY L
WILSON, TERRY
WILSON, TY E
WILTANGER, ERIC J
WIMBROUGH, JOHN H
WINCHESTER, BRIAN F
WINESKI, MARK A
WINFIELD, DWAYNE L
WISENER, JACK C
WOJCIEWEHOWSKI, THOMAS J
WOLF, BRUCE C
WOLFE, ERIC E
WOLFE, GARY M
WONG, FRED
WONG, RITCHELDA B P
WOOD, ERWIN
WOODS, CHRISTOPHER T
WOODS, FITZGERALD
WOODS, PAUL E
WOODY, RICHARD W
WOOLEY, JOEL M
WOOLFORK, IKE III
WOOTEN, TERRANCE D
WORKMAN, JOSEPH A
WORLEY, LONNIE
WORLEY, TIMOTHY A
WORTMAN, ARTHUR J
WRIGHT, LEVI
WRIGHT, WILLIAM E
WULICK, DANIEL III
WULK, STEVEN C
WUNDERLY, GEORGE R
WYNES, DAVID G
YAMBAO, ALBIN G
YAPTANGCO, MANUEL
YARBRO, MARVIN L
YARBROUGH, HORACE
YAUCH, ROBERT R
YCU, JAKE
YECKLEY, SHAWN R
YEWCIC, STEPHEN G
YI, DEBBIE R
YOUNG, BARON T
YOUNG, DENNIS E
YOUNG, GERALD E
YOUNG, KAREN P
YOUNG, RICHARD D
YOIJNG, ROBERT D
YOUNGER, WILLIAM R II
ZACH, ROGER A
ZACKERY, CHARLIE W
ZAHER, FRANCES JEANE
ZAHER, ROBERT
ZAPOLSKI, KENNETH S
ZARTMAN, MARK A
ZAYAS, ROBERT G
ZBRANEK, PETER J JR
ZEITERS, DAVID L
ZIEBERT, THOMAS J
ZIESMER, RICHARD J JR
ZIRTZLAFF, MARK T
ZUNIGA, JOSE L
ZWEIFEL, JOHN W
ZWICKER, JOHN J

Glossary

AAA	anti-aircraft artillery
AAFES	Army and Air Force Exchange Service
ABGD	Air Base Ground Defense
aerial port	a large area on the flightline where cargo was prepared to be loaded on transport aircraft and downloaded cargo was prepared for base distribution
AF	Air Force
AFP	Armed Forces of the Philippines
ALCE	airlift control element
American Condor	a ship designed to carry vehicles.
amplitude	relative strength
APO	air post office, a military zip code
ASAP	as soon as possible
barangays	villages
BDUs	battle dress uniforms
brick	hand-held radio. The nickname originated in the days when hand-held radios were the same size and weight as a brick.
BX	base exchange, similar to a department store
C3	command, control, and communications
CABCOM	Clark Air Base Command
CAT	crisis action team
CE	civil engineering
CEX	ration cards
CNN	Cable News Network
commissary	supermarket
COR	Condition of Readiness, a warning system used by military forces in the Pacific.
DOD	Department of Defense
EKG	electrocardiogram

Message boards at Travis AFB, California, for evacuees from the Philippines.

feeder band	thin band of clouds and rain preceding the main body of a typhoon.
FEN	Far East Network
frequency	pitch
GOP	the government of the Philippines
HF	high-frequency
humvees	high mobility multipurpose wheeled vehicles
ID	identification card
INP	Integrated National Police
jeepney	a popular form of transportation in the Philippines. Jeepneys were born from post-World War II military surplus Jeeps which locals bought at low prices and then extensively customized.
JTF-FV	Joint Task Force Fiery Vigil
JULLS	Joint Universal Lessons Learned System
K-loader	Equipment for loading cargo on transport aircraft.
lahars	mudflows
MAC	Military Airlift Command
MARS	Military Affiliate Radio System
MASS	military airlift support squadron
Medevac	medical evacuation aircraft
MHP	mounted horse patrol
MOGAS	motor gasoline
MREs	Meals Ready to Eat, the modern version of World War II Crations.
MSAs	munitions storage areas
MSSQ	mission support squadron
MWR	morale, welfare, and recreation
NCO	non-commissioned officer
NCOIC	non-commissioned officer in charge
NEO	non-combatant evacuation order
NPA	New People's Army
ORI	operational readiness inspection
PACAF	Pacific Air Forces
PAF	Philippine Air Force
PCS	permanent change of station
PDS	Personnel Data System
PF	pyroclastic flow
PHIVOLCS	Philippine Institute of Volcanology and Seismology (pronounced Fee-volks)
Plinian	explosive volcanoes
port call	scheduled departure

POV	privately owned vehicle
PPlan	programming plan
PVO	Pinatubo Volcano Observatory
pyroclastic	hot/fiery fragments
rack	bunk
ramp	aircraft parking area
Red Horse	Air Force combat engineering team
RSAM	relative seismic amplitude measurement
safe haven	an official designation under emergency evacuation orders, i.e. a place to live temporarily until a permanent assignment is tendered.
SAM	surface-to-air missile
sari-sari	primitive convenience store
SATO	scheduled Airline Ticket Office
sea-van	a tractor-trailer body that can be stacked until time for transport at which time it is lifted by a crane onto a tractor-trailer frame and driven to the port where it is then lifted onto a ship.
Shirt	squadron first sergeant
shopette	convenience store
SITREPs	situation reports
SOW	special operations wing
SP	security police
Stinger	small, heat-seeking missile carried in a short, lightweight launching tube
TDY	temporary duty
TFTG	tactical fighter training group
TFW	tactical fighter wing
Town Patrol	a special unit of the security police that patrolled the liberty areas surrounding Clark.
tropopause	boundary between the upper troposphere and the lower stratosphere; it varies in altitude from approximately 5 miles at the poles to approximately 11 miles at the equator.
UHF-SATCOM	ultra-high frequency satellite communications radios
USA	United States Army
USAF	United States Air Force
USAID	United States Agency for International Development
USCINCPAC	United States Commander in Chief, Pacific
USGS	United States Geological Survey
USMC	United States Marine Corps
USN	United States Navy
USO	United Services Organization
USPACOM	United States Pacific Command
VDU	visual display unit
water buffaloes	water carriers mounted on a small, two-wheel trailer
Wilco	standard radio terminology for "I will comply"

Bibliographic Notes

Situation Reports (SITREPs)

Every continuing military operation produces daily, weekly, or monthly situation reports. SITREPs keep everyone up the chain of command informed of the successes, failures, and needs of the troops who are engaged in action. During the Mount Pinatubo crisis, every crisis action team at every level of command produced them. The ones I focused on were the ones from the 3d Tactical Fighter Wing, Clark's main unit; the American Embassy in Manila; and the Joint Task Force-Fiery Vigil. Each of them produced a daily report that contains a wealth of statistical data about the volcano, actions at Clark, the evacuation, as well as commander's comments that paint a more personal view of events. SITREPs from the 3 TFW and Fiery Vigil are at the PACAF Command Historian's Office, Hickam Air Force Base, Hawaii. The only place I found copies of the embassy SITREPs was at US Pacific Command (USPACOM) headquarters at Camp Smith, Hawaii.

The Philippine Flyer

The *Philippine Flyer* was the official base newspaper at Clark Air Base. Before the Pinatubo eruptions it was a weekly, but immediately after the evacuation on June 10, 1991, the base Public Affairs Officer, Lt Col Ron Rand, directed a daily issue. He did so as a measure to contain the rumor mill active at an Air Force base, and at Clark the mill ground a lot of grain. Researchers unfamiliar with base newspapers should look at the issues with caution. Base newspapers are official publications; therefore, they tend to put an optimistic spin on all news. However, The *Philippine Flyer* provided me an invaluable source for information on people who were doing everyday jobs. Each edition of the "Magmatic Daily Tremblor [sic]," as it was called had a "Dust Buster of the Day," segment that honored some Ash Warrior who was doing a superlative job. These short features are the source for many whom I identified in the book. Many Ash Warriors kept each issue. I used a book-bound set that belongs to Ron Rand, currently the Public Affairs Officer of the Air Force. The PACAF history office has a set as well.

Interviews

I interviewed or surveyed nearly 100 people in the course of my research. With one exception, everyone I approached wanted to tell me his or her story. I tape recorded most of the interviews, and the tapes, unedited, are at the Air Force Historian's Office at Bolling Air Force Base, DC. Transcripts were made of some. Since my family and I were reluctant participants in the crisis, I thought I had a good feeling for what I would learn during the interviews. I was wrong. I was unprepared for the raw, deep emotions that came out of nearly every subject. As I mentioned in the book, the interviews were a catharsis for many, and it was not uncommon for tears to fall as some related their stories. Without question, the interviews are the most interesting parts of the story and give a rich resonance to the efforts of the Ash Warriors. Of course, interviews are like eyewitness accounts to a detective. Some are more reliable than others, but the emotion the interviews reveal is their real value.

An aerial view of Clark AB, with the base hospital in the center, after the eruptions.

Fire and Mud

This book, edited by Chris Newhall of USGS and Ray Punongbayan of PHIVOLCS and published by the University of Washington Press in Seattle, 1996, is the definitive scientific work on the Pinatubo eruptions. It is over 1,100 pages long and contains over 60 papers written on every aspect of the eruptions and subsequent lahars. Although most of the articles are beyond the comprehension of readers who are novice volcanologists, some deal with the warning system and evacuation coordination. For those who are especially scientifically-disadvantaged, the book contains scores of superb photographs of the volcano, the eruptions, and their aftermath. Certainly, no discussion of Mount Pinatubo or research of its eruptions would be complete without reference to this magnificent book.

JULLS Reports

The Joint Universal Lessons Learned Reporting System (JULLS) is a multi-service vehicle for recording lessons learned in joint operations. The JULLS reports for Joint Task Force-Fiery Vigil are all available at the PACAF Command Historian's Office, Hickam Air Force Base, Hawaii. The reports are formatted so that each page is an individual lesson learned from units that served in Fiery Vigil. The reports are very specific in that there is a page for even the seemingly most insignificant lesson. I used the reports as background for the narration. They are "must-read" for anyone researching Fiery Vigil.

Clark Air Base Closure Report

This report is included in its entirety in Appendix 1 because it contains lessons learned from each individual unit assigned to Clark Air Base. It was difficult to find, and I was lucky to stumble across it stapled to the back of the Clark Air Base closure plan, or PPlan that I found at the PACAF history office. Although I found most of the information credible in the report, a few of the accomplishments in the report are obviously exaggerated or overly optimistic. Readers will need to use caution when drawing conclusions from the report alone. The optimism of the report is not surprising. The Ash Warriors were writing it as they were in the final stages of closing the base and going home. It was an optimistic time, and one to reflect on tremendous successes. A few overdid it.

Joint Task Force-Fiery Vigil, 8 June-1 July 91

Anne Bazzell, now of the PACAF history office, wrote this fine accounting of the Fiery Vigil task force while she was the historian of the 834th Airlift Division, Hickam Air Force Base, Hawaii. It is a detailed report of the airlift effort for Fiery Vigil and contains comprehensive statistics and results of the Military Airlift Command's efforts in support of the task force. This report, together with the JULLS reports and the Fiery Vigil SITREPS, provides a comprehensive documentation of Fiery Vigil.

Photographs and Videotapes

Nearly every Ash Warrior has an extensive collection of pictures and videos from the disaster. The base librarian at Clark was a Filipina who was still there after the smoke cleared. She sponsored a photo contest in the library as a diversion for the Ash Warriors. Some submissions were nothing short of spectacular and several are included in this book. Many Ash Warriors have personal collections of the contest winners, and most Ash Warriors have extensive photos they took themselves. Combat photographers were at Clark throughout the ordeal, so a bonanza of pictures, both still and moving, is available at the Defense Visual Information Center. It is likely that every person who was at Clark or evacuated from Clark has either one or both of two excellent television presentations that were shown on national networks. One was produced by Nova and the other by National Geographic.